Eclipses 2005–2017

Eclipses 2005–2017

A Handbook of Solar and Lunar Eclipses and Other Rare Astronomical Events

Wolfgang Held

Floris Books

Translated by Christian von Arnim
Edited by Christian Maclean

First published in German as *Astronomische Sternstunden*
by Verlag Freies Geistesleben in 2005
First published in English by Floris Books 2005

Previous page: The sun's corona visible during a total eclipse

British Library CIP Data available

ISBN 0–86315-478-6

Produced by Polskabook, Poland

Contents

Foreword

There are moments in which we are filled with such a wealth of impressions, feelings and thoughts that time seems to expand to such an extent that it almost seems to stop. A little bit of eternity seems to come within our grasp at such special times.

Everyone has experienced such distinct moments. It may be a human encounter which grows into lifelong friendship, the experience of the birth of a human being, a phenomenon of nature. But there are also spell-bound moments which take hold of us unnoticed by the people around us. New ideas, new confidence, a victory over oneself or the rare peace of reflection. In spell-bound moments we experience more of nature, experience its deeper layers. In personal spell-bound moments we come to a deeper encounter with ourselves in a similar way.

Many such hours of inner peace are heralded when we look up at the stars of the night sky. For all the hectic bustle and arbitrary events in our lives, our gaze upwards reaffirms the certainty in us that there is a meaningful relationship, a deeper cohesion between the environment and ourselves. Such a feeling, when faced with the tranquil constellations, raises questions in us about the really important things in life. If we only ask such questions when the mortgage is no longer a problem and our career is assured, it is often too late. We can learn a lot from the starry sky in this respect — as I hope this book makes clear — down into our very personal concerns.

It might sound paradoxical, but the infinite expanse and depth of the cosmos can guide us towards our most personal and intimate concerns. The aspect of the starry heavens not only gives us expanse and eternity, but also the kind of inwardness which otherwise only comes to expression in the human soul. Of course the modern world view of the natural sciences pushes such feelings aside as soon as they appear. But they should be taken seriously for they lead us to what we might call the soul of the cosmos.

But alongside these personal astronomical spell-bound moments, there are also those which affect not just the individual but all of us together. These are moments in which the cosmos appears to celebrate its own existence. Time seems to expand during the majestic spectacle of a solar eclipse or the tranquil poetry of a gathering of the planets. To point to such moments, to prepare for them so that when they occur in the heavens we can forget everything except our attentive, wakeful observation — that is the purpose of this book.

It belongs to the natural character of cosmic events that there is no optimum point from which they can be observed. They are far too grand to be capable of observation in their full scope from a single chosen spot. It is nevertheless worthwhile to take account of the various meteorological conditions when we want to observe a solar eclipse. Hence the descriptions of the individual eclipses contain notes about the local probability of finding a clear sky. Should you still have bad luck and find that a cloudy sky restricts visibility, then the experience of many people who travel to eclipses may be of consolation that a solar eclipse is such a mighty spectacle, it transcends physical conditions with such grandeur, that it makes its imprint as an unforgettable event whatever the weather, be it cloudy or even pouring with rain.

Eclipses take place usually twice a year, but we have to be in the right location to see them. Over the next few years not only will a total solar eclipse pass over the one of the remotest islands, Easter Island, in 2010, but others will cast their shadow over the monumental Grand Canyon, the elemental nature of the remote southern Andes, the tropics of Indonesia or the icy landscape of Spitsbergen. Holiday areas like the Maldives and Réunion or places imbued with history such as eastern Turkey will turn into destinations for travellers wanting to see the eclipse. The path of the moon's shadow will pass over many unique places on Earth in the next few years so that a journey to those locations will be a profound experience. This book is an invitation to participate in the cosmic events of the coming years.

I would like to thank Christian Maclean for the idea of this book, and for his suggestions and encouragement in writing it.

Dornach, Switzerland, Summer 2005

Three Ways of Gaining a Relationship to the Cosmos

In the spring of 2002 there was an exceedingly rare and impressive gathering of the planets. All the wandering celestial bodies visible to the naked eye were positioned in a line rising in the spring evening sky. No one who became aware of this planetary gathering could deny the magic of the heavenly phenomenon during the Easter period. Self-observation shows in this as in other cases that it does not take much to transform the wonder at the magnificence of such a natural spectacle, the string of light stretching from Mercury on the horizon to Jupiter at the zenith, into a certain fear. It is today's concepts in astronomy and astrophysics of infinitely large space in which the Earth becomes vanishingly small which then arise in the soul and put an end to such an unselfconscious view. There is a change in the feeling of standing in the heavenly vault, the feeling of quiet solemnity in which all talk ceases, as can be experienced among a group of people observing the heavens. Faced with concepts of clusters of galaxies and light which takes millions of years to reach Earth, the original feeling of homeliness changes into one of being lost.

In such moments we become conscious of something which is normally overlaid by the concerns of our daily lives: the idea of the Earth as a speck of dust in the cosmos. The significance of this image should not be underestimated. Everything which gains in importance through inner endeavour of the soul, which we want to take seriously with all our heart, becomes meaningless in the face of this idea. Where we seek the validity of an absolute, such imposing astrophysical ideas blend a delicate relativism with our own feelings.

Even if we allow the Earth as our home planet a special place in the cosmos, today's materialistic view nevertheless remains unsatisfactory with regard to the feelings which arise when we allow ourselves to be influenced by the heavenly vault at night.

We feel that astrophysics, which is only concerned with material connections in the cosmos, cannot grasp the whole of reality.

Astrophysics, the science of the lifeless part of the cosmos, must be combined with a science which tries to understand the life, the soul and the spirit in the cosmos. The pronounced longing for such a direction in research is currently expressed in the wealth of literature about spiritual astronomy and cosmology. It was not surprising to read in the German weekly newspaper *Die Zeit* some years ago that a non-fiction best seller must contain three words in its title: God, spirit and cosmos.

There are many impressive descriptions by astronauts about their feelings during their flight in space. They relate largely to Earth and not to the grandiose cosmos which surrounds them. Why? Why are astronauts gripped by a love for the Earth which almost goes beyond what is human when they observe the planet from the window of their capsule? They do so because the scientific view shows them a cosmos which they experience as hostile to life in all its manifestations, and the Earth is suspended in the midst of this life-threatening environment. Although the Sun supports life on Earth as part of the extra-terrestrial cosmos, the scientific view nevertheless creates the certainty that the cosmos which surrounds the Earth denies life. From a modern point of view, the Earth in the cosmos appears like a nutshell filled with life in an ocean of death, and the question might well arise as to why such a hostile cosmos tolerates the Earth and its infinite richness in life forms at all.

Such a view of the world, restricted to the physical, in which the development of life, of ensouled and, finally, human consciousness, appears almost like an occupational hazard, has a bearing on our actions: every deed which is motivated by the wish to contribute to giving meaning to the connection between all life forms, be it in an ecological or social sense, must exclude the life-denying cosmos in such a situation. The much-used term 'holistic' must always remain relative against the background of a cosmos which is seen in material terms because everything which happens on Earth takes place in a cosmic context which appears to be totally unrelated.

In order to recover or develop an understanding of the cosmos which reckons with life — be it completely non-earthly — and a spiritual element in the cosmos, three paths are necessary. These paths can be summarized under the headings *observation, interdisciplinary research* and *meditation.*

Observation here means that we learn to observe the natural phenomena with such attention and lack of preconceptions that they increasingly reveal their true being. In a human face, we find it easy to recognize the current mood of soul by the physical expression. Here the spiritual element is expressed in a form we can understand through the facial physiognomy and expression. Enhanced attention and concentrated lack of preconception are necessary to grasp the being of external nature, of plants and animals right up to the planetary constellations. Just as in understanding human beings our prejudices ought to be repressed, so in the observation of nature our theoretical knowledge should for a time be left behind. Such a concentration on the actual phenomena is called phenomenology or Goetheanism. In geography and biology, in particular, this research method has produced wonderful findings about the characteristics of natural phenomena.

In the past thirty years, our knowledge of the relationships between Earth and the cosmos has advanced in leaps and bounds. Be it the relationship of the water balance in trees to the phases of the Moon, the pulsating of oak buds in correlation to Mars and the Moon or the cosmologically-related rhythms of the human organism: the close relationship between nature and, in a certain sense, human beings to the cosmic environment is becoming more and more clear. New research fields such as cosmological botany, cosmological anthropology and cosmological chronobiology are producing exceedingly interesting new perspectives and ways of understanding ourselves and surrounding nature. By learning to understand nature better through such *comparative studies,* we are also gaining a deeper understanding of our cosmological environment, the planetary system.

Anyone who tries to understand the cosmos in spiritual terms soon encounters the fact that to look at the infinity of the stars is closely related to immersion into our own inner being. The fixed relationship of the stars to one another, which we can perceive as a pure manifestation, corresponds to the life of our thoughts and ideas. Not for nothing do we refer to the cosmos of ideas. Just as concepts are not subject to our whims, so the stars cannot be manipulated. Although we observe them with our physical senses, they cannot be grasped — they are at the boundary of the physical. The Latin verb *considerare* (to consider) is characteristic of this relationship between thinking and the stars. Translated literally, it means 'to be under the stars.' Just as in English usage the 'elevation' of a matter or a thought is almost synonymous with its 'profundity,' so the inner personal world as can be accessed through *meditation* and the expanse of the cosmos are very close.* Immanuel Kant expressed it in his much quoted sentence:

> Two things fill the mind with ever new and increasing wonder and awe the more and continuously we reflect on them: the starry heavens above me and the moral law within me.

In the present book we will embark on the first of these three paths, the path of observation. We will focus on exceptional astronomical events such as solar and lunar eclipses, as well as notable constellations, because they express the spiritual life of the cosmos in an enhanced, and sometimes even overwhelming form.

But since the solar eclipses which provide the focus of this book occur all over the Earth, this book is also a kind of travel guide to all the places where the eclipses can be observed, with special value having been placed on precise locations and times as well as detailed maps. The author hopes in this way to give all travellers who follow the solar eclipses a valuable tool in their search to come one step closer to the cosmos.

* See, for instance, *The Dreamsong of Olaf Åsteson ,* Dante's *Divine Comedy,* or *The Anticlaudian* by Alanus ab Insulis.

The Miracle of a Solar Eclipse

As when the Sun, new risen,
Looks through the horizontal misty air,
Shorn of his beams, or from behind the Moon,
In dim eclipse, disastrous twilight sheds
On half the nations, and with fear of change
Perplexes monarchs.

Milton, *Paradise Lost,* 1667

Storms at sea as well as earthquakes and volcanic eruptions are probably the only other natural phenomena which affect people like a total solar eclipse, never to be forgotten; and it affects them in such a way although no threat whatsoever emanates from a solar eclipse. Calculated a long time in advance with plenty of notice, the celestial spectacle takes place and yet people are still shocked and deeply touched by the silent immensity of the Sun being covered by darkness. We ourselves become as quiet and speechless as nature surrounding us: no dog barks, no birds sing when the greenish-grey darkness of the core shadow of the Moon falls across the landscape and the sky becomes so dark that planets and bright stars become visible during the day.

One would have to be a poet in order to be able to express the strong and changing feelings provoked by a solar eclipse, which is why the words of Wordsworth and the report of the last visible solar eclipse in Central Europe by Adalbert Stifter is included in this book. Anyone who has ever experienced a solar eclipse will see no pathos in the words of the Austrian writer.

It is easy to understand that in earlier times people were filled with fear and terror by a solar eclipse, because in ancient Greece, for example, only a few educated people understood the astronomical circumstances leading to an eclipse. The 'theft' of the sunlight by the Moon suddenly called into question the goodness and divinity accorded to the Sun in religious experience. The Moon turned into a dragon which devours the Sun — that is how many ancient cultures thought about the eclipse. Yet why are we still affected just as strongly today, in a time when everyone understands the astronomical details to a greater or lesser extent and everything can be explained?

It is because, in the same way as an earthquake, the secure foundation of our existence is suddenly called into question for a short period of time, in this instance, the certainty that the Sun shines during the day. We may object that it simply becomes night for a brief period of time, but that is to miss the point. The kind of darkness which arises during a solar eclipse is not the kind of darkness we know from night, it has nothing of the romantic mood of a beautiful dusk. It is a greenish light, sallow and strangely oppressive, in which the now visible veil-like ring of rays of the Sun's corona thrones majestically above this lowering atmosphere.

Observation of the eclipse

A suitable location for observation

In order to observe an eclipse, we should try to position ourselves in the line of the core shadow, the umbra, since the total eclipse of the Sun is far more spectacular than a partial eclipse south or north of this corridor of total shadow. (Anticipate possible traffic jams in populated areas).

Although a total eclipse mostly takes longer than two minutes, it seems to be over with great speed. Hence we should be as close as possible to the central line since the duration of darkness rapidly reduced to the north and south. A place should be chosen which has an unrestricted view of the horizon, above all the western horizon, for which raised ground giving a view over the western landscape is best. From such a location one can already see on the horizon the impressive approach of the eclipse during the time when the Moon has not yet fully covered the Sun. A peculiar darkness, like in a distant thunder storm, approaches with ever greater speed from the west.

Sufficient peace for the celestial spectacle

In order to be able to observe the many different events in the sky during the eclipse, and also to be able to see how nature responds to the celestial spectacle, it is helpful if the selected location is sufficiently peaceful. Often cities falling within the eclipse offer many different events. Large projection screens are set up on which the approach of the eclipse is shown with a loudspeaker commentary. But to really experience the eclipse a quiet place without diversions, out in the countryside if possible, is to be recommended. If we are in a group of people conversation will die away as the total eclipse begins, and we can completely abandon ourselves to its mighty impression.

The drama of the course of the eclipse

With our eyes protected by a filter (for instance the glasses supplied with this book), the first impression we have when we suddenly see the slight dent at the edge of the Sun is probably not much different to Tycho Brahe's impression who observed a solar eclipse in his youth: the precision with which the Moon starts to move in front of the Sun at the second which has been calculated in advance is deeply impressive. It appears to be like an inexorable meeeting of fate.

The surroundings do not yet appear to grow dark because even when 50% of the Sun has been covered the eye compensates for the fading light, but a dramatic change has nevertheless taken place. Butterflies disappear, swallows fly close to the ground in agitation, the mood in nature becomes more and more depressed. A dull greenish, mustard-coloured yellow settles on the fields and we feel strangely disorientated the nearer we approach totality.

About five minutes before the end of the partial phase events take on a dramatic turn: the sky becomes an indescribable greenish and sallow tint, which is never quite the same in different eclipses. It is quite distinct from cloud cover or normal dusk. The spreading shadow can be seen on the western horizon. Like a wave of darkness it takes hold of the clouds and appears to rise from the ground. It is the shadow of the Moon racing towards the observer at a speed of about 1 km per second. At the same time it begins to get cooler and any wind often dies down. The situation has a curiously unreal effect because opposing phenomena suddenly come together: on the one hand the level of light is similar to the advanced stage of dusk but at the same time one can see that shadows are etched very sharply, particularly on people's faces.

A few seconds before the Sun is fully covered its last rays of light fall on the Moon through the latter's valleys. The extremely thin solar sickle can then look like a necklace of pearls, producing a so-called diamond ring of light before the corona of the Sun becomes visible. A few minutes before totality, peculiar fleeting shadows, the so-called 'flying shadows' can be seen on the light surface. The speed of these mysterious dark bands which are about the width of a hand increases as totality approaches. After the last sunlight which has still found its way to Earth through the valleys of the Moon has disappeared, the pink chromosphere, a thin sheath of light surrounding the Sun, is visible for a few seconds until the Moon covers it too.

Now darkness closes in with a greenish and sallow light, comparable to a Full Moon night. Often groups of people fall into awe-struck silence. The birds grow silent, some blossoms close. The horizon has taken on a pronounced, eerie orange-red hue of dusk. Dominating everything, the Sun's corona is now visible: a white ring of rays with a bright inner part and a weaker outer part which disperses the further it extends from the Sun. The naked eye sees much more of the curved structures of this 'Sun atmosphere,' 'the Sun outside the Sun,' than can be captured in a photograph. The difference in brightness between the inner and outer corona exceeds the capabilities of a camera.

The colours in nature cannot be photographed either. They are still dimly there but it is as if nature has lost its soul.

In many familiar photographs, the prominences at the edge of the Sun look like a mighty fiery phenomena. But the actual experience during a solar eclipse tells a different story: the orange and red fringe of light exudes such mildness that we gain the impression that it is not the Moon which is victorious over the Sun, but that the Sun tolerates its own eclipse.

Much too fast, after a few minutes the light of the diamond ring appears again to the right* of the Moon shadow. With the first ray of the Sun, light and life return and we can experience a wonderful joy, a feeling of spring in nature as can only be described by poets. After a few moments, the Moon once again releases a delicate solar sickle. Now the phenomena occur in

* In the southern hemisphere the appearance is mirrored and the light appears on the left.

reverse order but without the ominous drama of events before totality: once again the flying shadows can be observed. The shadow of the Moon over the landscape disappears across the landscape towards the east at great speed, nature returns to life. Rapidly the surroundings grow light again. The temperature, which can easily fall by as much as 7°C, also rises again.

Just as we really learn to appreciate friends or our familiar everyday surroundings through their sudden absence, so it is with our relationship to the Sun. The experience of the eclipse turns our personal relationship to the Sun into something new, something more meaningful.

Protecting the eyes

During the partial phase, the Sun should only be observed with protected eyes, otherwise damage to the retina could occur. Normally the eyelids protect us from too much sunlight by blinking. But when looking at darkness in a concentrated way, this reflex is weakened. Even when only a small sickle of the Sun is left, too much light and heat penetrates the eye and can cause permanent damage to the retina. The impression of the weak light of the remaining solar sickle is deceptive. Since the retina cannot experience pain, we are not warned of permanent visual damage which may in any case only occur a few hours later. The most suitable filters are of the type enclosed with this book in the form of protective glasses made of aluminized mylar film or welding filter glass. Glass panes covered in soot, black film, or CDs block the visible sunlight but not the UV radiation and are dangerous. If a telescope or binoculars are used to observe the eclipse, a corresponding mylar film must be attached securely to the front of the instrument, not between the eye and the lens. If children are around, the telescope must be under constant supervision for as long as it is pointed at the Sun.

No filter is, of course, used during the period of totality.

Projection is another possibility for observing the solar sickle. Without a filter, the sunlight is allowed to fall on a white sheet through binoculars on a stand. A piece of cardboard with a small hole pierced in the middle could even be used instead of binoculars.

A simple but slightly more comfortable form of projection is provided through the construction of a camera obscura: take two boxes, of which one is slightly larger than the other, so that the smaller one can be pushed into the larger one (once the ends have been removed). A small hole should be pierced at one end while the cardboard is cut away at the other and replaced by greaseproof paper as a matt screen.

This type of projection can be found a million times over in nature: if we look at a rhododendron bush, for example, or at the shadow of a deciduous tree we can often see hundreds of superimposed sickles of the Sun on the ground. The sunlight breaks in the cracks between the individual leaves just like in the pinhole.

A piece of cardboard with a small hole pierced in the middle can be used to project an image of the Sun.

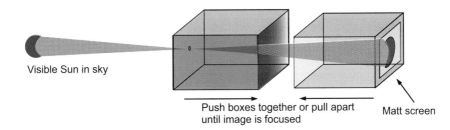

Visible Sun in sky

Push boxes together or pull apart until image is focused

Matt screen

Two boxes, one being slightly larger than the other, so that the smaller one can be pushed into the larger one (once the ends have been removed). Pierce a small hole at one end. At the opposite end cut away the cardboard and place a sheet of greaseproof paper as a matt screen.

How do Solar Eclipses Occur?

The course of the Moon and the Earth

The different phases of the Moon arise through the trajectory of the satellite around the Earth. If the Moon is next to the Sun we have New Moon, if it is positioned at right angles to the Sun we have the waxing and waning half Moon and at Full Moon it is positioned opposite the Sun. The designation 'New Moon' is, however, misleading because the Moon is not yet 'new.' In fact, in antiquity the New Moon designated the small lunar sickle when the Moon was visible again for the first time in the evening sky

The phases of the Moon

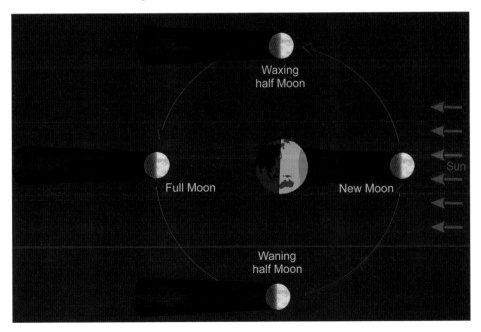

as a New Moon. The Roman priests proclaimed the new month when they saw the New Moon. The word calendar is reminiscent of this, since *calere* means 'to call.' In this respect the New Moon should actually be called the non-moon.

Total and annular solar eclipses

A solar eclipse occurs when the shadow of the Moon falls on the Earth. Now each body when it is illuminated by a light source which is not a point throws two different shadows — the core shadow or umbra, and the penumbra or half shadow. If we are within the core shadow we see the Sun completely covered from this perspective. It is the range of a total solar eclipse which can be up to 270 km wide. The movement of the Moon means that in the course of an eclipse the umbra sweeps a path of darkness on the Earth. Outside this core zone, in the range of penumbra, the Sun is seen partially covered by the Moon. The diameter of the penumbra is approximately 7000 km.

It is one of the mysteries of the planetary system that the Sun and the Moon display correlations which can hardly be coincidence: it is indeed astonishing that the Moon with its 29.5 days from Full Moon to Full Moon takes almost the same amount of time as the Sun requires on average for one rotation around its axis. The Moon thus reflects not only the light of the Sun but also its type of movement. The second correlation between Sun and Moon lies in their size in the sky: although the Sun with a diameter of 1 390 000 km is approximately 400 times as big as the Moon at 3476 km, both appear to be the same size when looked at from Earth. The Moon with its average distance from the Earth of 384 000 km is thus 400 times closer than the Sun.

However, the distance between the Moon and the Earth fluctuates due to the elliptical path of the Moon around the Earth. At apogee (furthest from Earth), the distance is about 356 410 km and at perigee (closest to Earth) about 406 700 km. This means that the distance of the Moon moves by one seventh. It is worth trying to visualize this difference in size by comparing the Full Moon when close to and distant from the Earth. Since the height of the Moon above the horizon influences the impression of size, the Moon should stand as high as possible. The problem is, that the Full Moons at perigee and apogee lie approximately half a year apart. Nevertheless, we may become more aware that the summer Full Moons in the next few years are particularly large and the winter and spring Full Moons small.

The table shows some examples in the next few years.

Full Moon and perigee less than 2 days apart = large Full Moon	Full Moon and apogee less than 2 days apart = small Full Moon
2005 July 21	2006 Feb 13/14
2005 Aug 19	2006 March 13/15
2005 Sep 16/18	2007 April 2/3
2006 Aug 9/10	2007 April 30/May 2
2006 Sep 7/8	2008 May 20
2006 Oct 6/7	2008 June 16/18
2007 Sep 26/27	2009 July 7
2007 Oct 26	2009 Aug 4/6

The times of smaller and larger Full Moons move forward each year by about one month. This is because the location of perigee and apogee on average move forward by about 3.1° and correspondingly pass through the zodiac once every 8.75 years.

But it is not only the Moon which fluctuates in its distance and thus in its size when seen from the Earth. Since the Earth moves in an ellipse around the Sun, even if it is less elongated, the distance between the two and thus the apparent size of the Sun also change. The distance between Sun and Earth ranges between 147 and 151 million km in the course of the year, so that correspondingly the apparent diameter of the Sun changes by 3%. This fluctuation is four times smaller than the fluctuation in diameter of the Moon, but it contributes to the complexity of time and character of solar eclipses.

In optimum conditions, that is when the Moon is at perigee and the Sun at apogee, the tip of the umbra cone projects an umbra which is 270 km wide. In the opposite case, when the Sun is large (perigee) and the Moon is relatively small (apogee), the shadow cone does not reach the surface of the Earth. Instead of a total eclipse, an annular one occurs in which the Moon is surrounded by a ring of light. In such a case the remaining brightness is so great that the corona described above cannot be seen, but the ring of light around the Moon, which shimmers in an anthracite colour, is still a breathtaking sight.

A special case occurs when the relationships of size between Sun and Moon are such that the umbra of the Moon only just reaches the Earth's surface. Then an annular eclipse occurs at the start and end of the eclipse, when the shadow has to travel a longer path to the Earth's surface, and totality occurs in the middle of the eclipse. This is a hybrid eclipse.

The rhythm of solar eclipses

A solar eclipse occurs when the Moon casts its shadow on to the Earth, but why does this not happen at every New Moon when the Moon is between the Earth and the Sun? The path of the Moon is inclined by 5° to the Earth's. The angle of its path means that the Moon as a rule travels above or below the Sun as it passes. As a consequence its shadow mostly does not fall on the Earth but passes above or below it as the case may be.

The Moon's path must cross the ecliptic at New Moon for a solar eclipse to occur.

Two things must thus come together in order for a solar eclipse to occur. There must be a New Moon and at the same time this New Moon must be at the same 'level' as the Earth and the Sun; in its course it must break through the plane of the Earth's trajectory, the ecliptic. These breakthrough points of the Moon's

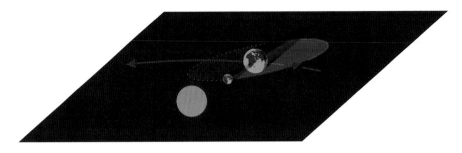

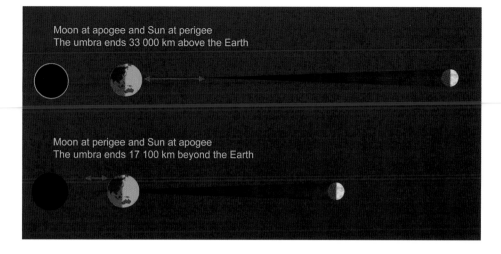

Moon at apogee and Sun at perigee
The umbra ends 33 000 km above the Earth

Moon at perigee and Sun at apogee
The umbra ends 17 100 km beyond the Earth

trajectory, lying opposite one another, are called the lunar nodes. Only when the line connecting these two points is on a level with the Sun can an eclipse occur. By its nature, that happens twice every year. On the one occasion the Sun is aligned with the ascending node and, correspondingly, half a year later in line with the descending node. However, the New Moon need not necessarily reach this position in relation to the Sun to the exact day in order for a solar eclipse to take place.

Since the Earth is clearly larger than the Moon, the latter's umbra falls on the Earth even if the Moon is positioned as a New Moon between the Sun and the Earth up to 18 days before or after the ideal Sun position. Now this timeframe of 2 x 18 = 37 days is greater than the rhythm of the moon's orbit. New Moon occurs every 29.5 days. Hence a solar eclipse always takes place in the periods in which the Sun stands near the nodal line. So at least two and a maximum of four eclipses take place per year. As already described, they can be either central (total or annular) or partial.

We will explain this in detail using the example of the solar eclipse of February 7, 2008: the Sun reaches the ascending lunar node on February, 16, 9 days after New Moon. On the day of New Moon, February, 7, the Moon is thus positioned slightly below the position of the Sun so that the umbra falls near the South Pole.

The distance of the Moon to the Earth influences the size of the shadow falling on the Earth.

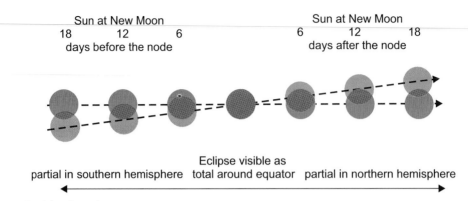

Sun at New Moon
18   12   6
days before the node

Sun at New Moon
6   12   18
days after the node

Eclipse visible as
partial in southern hemisphere   total around equator   partial in northern hemisphere

Partial and cental eclipses with New Moon before and after a node.

The movement of the nodes

It is typical of the planetary system that almost any motion has a countermotion. There is a whole canon of movements and counter-movements in the planetary system. This also applies to the Moon: while the Moon travels its path around the Earth from west to east, the nodes move in the opposite direction from east to west. The plane of the moon's path thus behaves like a plate which has been spun on a table at an angle and which rotates tottering from side to side.

The lunar nodes move about 20° per year so that the solar eclipses take place on average three weeks earlier each year. The shift of eclipse periods in the next few years is shown in the table.

Eclipse at the ascending node	Eclipse at the descending node
2005 April 8	2005 Oct 3
2006 March 29	2006 Sep 22
2007 March 19	2007 Sep 11
2008 Feb 7	2008 Aug 1
2009 Jan 26	2009 July 22
2010 Jan 15	2010 July 11
2011 Jan 4	2011 June 1, July 1

Saros periods — the 42 eclipse series

It takes 18 years 11 days and 8 hours for the nodes to move completely round the zodiac and for almost identical conditions resulting to recur in another similar solar eclipse. This period is called Saros period. The shift of eight hours, however, means that the Earth turns by a third so that a different area is covered by the shadow zone. Only the fourth eclipse in the series, after 54 years and 34 days, will reach the same area, though it will occur about 300 km further north with an ascending node or a similar distance south with a descending node.

Each eclipse thus belongs to its own series of eclipses which uniformly pass across the Earth in a period of about 1000–1200 years. Such a series of eclipses starts at the ascending node as a partial eclipse in Antarctica. The umbra does not yet touch the surface of the Earth. Only after 10 eclipses or 200 years, has the eclipse 'grown' so far that the central shadow touches the Earth. There now follow 900 to 1000 years of total (or annular) eclipses until finally the umbra leaves the Earth again in the Arctic and the series concludes with ten to eleven partial eclipses. There are 42 different series of such eclipse groups all of which characteristically mark the Earth with shadow lines in a cycle of 18 years and 11 days and which jointly produce the average of 2.3 solar eclipses per year. Each series has its own character. When such a Saros cycle has come to an end, it is replaced by a new one.

In contrast to the continual flow of light and heat from the Sun to the Earth's surface, the group of 42 eclipse series etch lines of shadow into the Earth's surface.

The sensory-supersensory nature of an eclipse

Most people who try to describe the phenomena and impressions arising from a solar eclipse reach the limits of their own descriptive capabilities. One has to be a poet to be able to do descriptive justice to the might of such a natural spectacle. Alongside the difficulty of being able to grasp the enormity of a solar eclipse, descriptions generally resemble one another in being 'simultaneously terrible and sublime.'

There is probably no other spectacle of nature, no other impression which combines grandeur and oblivion, which is simultaneously sublime and terrible to such a degree. On the one hand, we experience colours of unfamiliar ugliness in the yellow, greenish, mustard hues of the landscape during the period of the eclipse. Anyone trained to sense the inherent quality in colours is struck dumb by the dark, almost diabolical character of the tone of the colours while the Sun is eclipsed. At the same time the Sun's corona is enthroned above this dismal scene, the tenuousness and transparency of which were rightly described by Adalbert Stifter as 'the most fascinating light effect I have ever seen,' or by Wilhelm Meyer as 'the mysterious light from the other side.' Binoculars or a telescope allow us to see the photosphere and solar flares. From photographs and physics we are used to thinking of them as related to fire. The view through the telescope reveals a surprisingly different characteristic: the outer layer, the shining, rose-coloured ring around the Sun possesses a transcendental delicate nature which no camera is able to capture.

The dual nature we have described is reminiscent of the character of supersensory experience. All such descriptions as well as personal biographical experience possess the characteristic that they contain both grandeur and oblivion. 'Every angel is terrible,' the poet Rainer Maria Rilke wrote at the start of the *Duino Elegies,* setting out this contradiction. But this is not the only view placing such a natural spectacle close to supersensory experience. Two further aspects underline the affinity of this phenomenon with the experience of supersensory reality: 'It was as if nature had been fractured. The pulses of physical nature faltered, nature itself seemed to stop in its tracks.' (Wilhelm Meyer, the Egyptian eclipse of 1909, see p. 139)

Part of the experience of a solar eclipse is that our feeling of time is completely transformed. Whereas the time of the eclipse appears to have passed in no time at all when we look back on it, it seems to last an eternity during the actual experience. But this changed sense of time is felt as being so real during the eclipse that it is experienced as a disturbance in time of nature as a whole as described in Meyer's quote.

'I felt as if I was on a different planet during the eclipse,' a Swiss school student reported after the 1999 eclipse.

Such descriptions need to be taken literally if we want to grasp the inner character of a solar eclipse. As the light of the Sun disappears, we curiously also lose the Earth along with the sunlight, and we appear to be removed from physical reality. It is an experience which tells us that existence on Earth in a comprehensive sense is simultaneously existence *in* the sunlight. If the sunlight disappears not through the natural rhythm of day and night, we feel that our normal physical existence has ceased to have legitimacy.

These three characteristics — the dual nature, the temporal fracture and the alienation from the Earth — belong to the central feelings which pass through the soul in the experience of a total solar eclipse. They place the spectacle in the heavens close to supersensory experience, and show that what reason ascribes to the geometric constellation of far distant heavenly bodies takes place, in terms of its spiritual content with equal reality in our own soul.

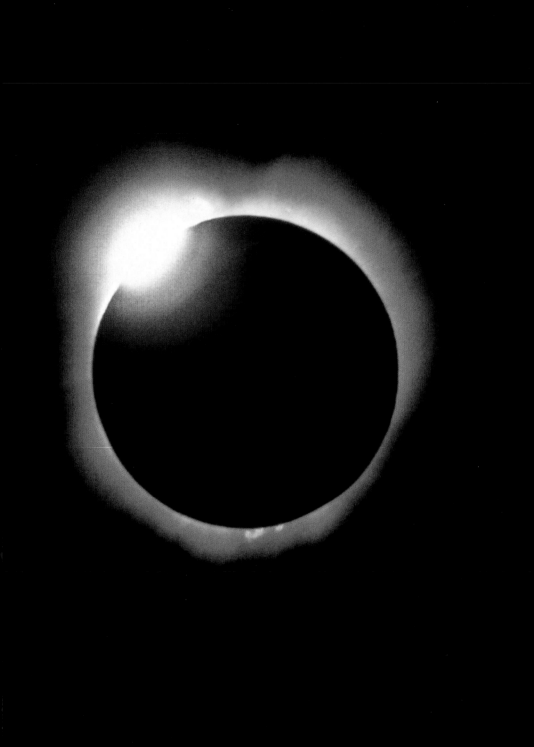

Solar Eclipses 2005–2017

Date	Place	Type	Page
2005 Oct 3	Portugal, Spain, North & East Africa, Seychelles	Annular	35
2006 March 29	Brazil, West Africa, Egypt, Turkey, Central Asia	Total	41
2006 Sep 22	Surinam, French Guyana, Brazil, Atlantic	Total	48
2007 March 19		*Partial*	
2007 Sep 11		*Partial*	
2008 Feb 7	Antarctica	Total	51
2008 Aug 1	Canada, Greenland, Russia, Mongolia, China	Total	54
2009 Jan 26	Indonesia	Annular	59
2009 July 22	India, Nepal, Bangladesh, China, Japan, Kiribati	Total	62
2010 Jan 15	Central Africa, Kenya, Sri Lanka, Burma, China	Annular	70
2010 July 11	South Pacific, Easter Island, Chile	Total	77
2011 Jan 4		*Partial*	
2011 June 1		*Partial*	
2011 July 1		*Partial*	
2011 Nov 25		*Partial*	
2012 May 20/21	China, Japan, North Pacific, United States	Annular	82
2012 Nov 13/14	Australia, South Pacific,	Total	90
2013 May 10	Australia, Papua New Guinea, Polynesia, Kiribati	Annular	93
2013 Nov 3	Atlantic, Gabon, Congos, Uganda, Kenya	Hybrid	98
2014 April 29	Antarctica	Annular	102
2014 Oct 23		*Partial*	
2015 March 20	Faeroes, Spitsbergen, Arctic Ocean	Total	104
2016 March 9	Indonesia, Pacific	Total	108
2016 Sep 1	Congos, Tanzania, Madagascar, Réunion	Annular	112
2017 Feb 26	Chile, Argentina, Angola	Annular	116
2017 Aug 21	United States of America, Atlantic	Total	120

Previous page:
A diamond ring
visible just before
complete totality

2005 October 3, annular

Annular: Portugal, Spain, North and East Africa.
Partial: Europe, Africa and Middle East, Indian subcontinent.

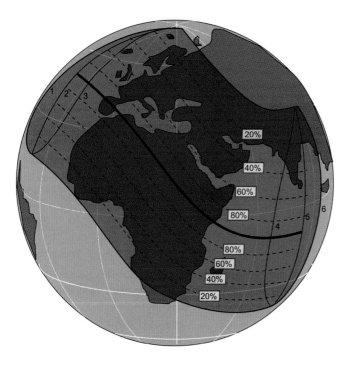

Course of the eclipse

1 End of eclipse at sunrise

2 Maximum at sunrise

3 Beginning of eclipse at sunrise

4 End of eclipse at sunset

5 Maximum at sunset

6 Beginning of eclipse at sunset

On September 28, the Moon is at apogee. Five days later it casts its shadow on the Earth. The proximity of the solar eclipse to the apogee means that the cone-shaped umbra ends before it reaches the Earth. Seen from the perspective of Earth, only 90% of the Sun is covered in the central range of the eclipse. The remaining

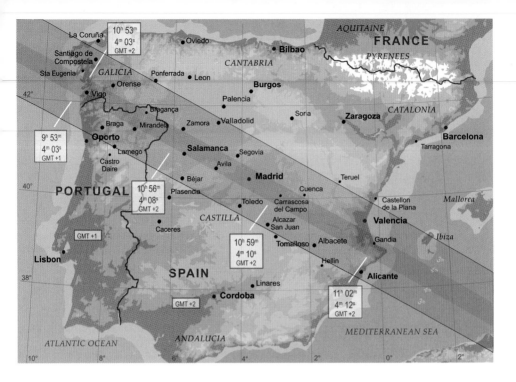

disk of the Sun triumphs as an indescribably imposing solar ring around the anthracite shimmer of the darkened Moon.

The band of the eclipse moves through northern Portugal and across Spain. It moves first through Galicia, sweeping across the pilgrimage site of Santiago de Compostela at a speed of 6 000 km/h.

Salamanca, Braga, Madrid and Valencia also lie in the eclipse zone. For more than four minutes at a time, 90% of the Sun is covered from the peninsula, so that only a circular fringe of light remains around the anthracite coloured Moon.

The shadow, which soared across the Atlantic at 10 000 km/h, slows down on the Iberian peninsula to about 5 000 km/h; the length of the eclipse in the 180 km-wide band also grows to more than 4 minutes. Ibiza, the southernmost island of the Balearics is grazed, although there, the eclipse only lasts for 2 minutes. The path of the eclipse reaches land again at Algiers at 09:08, travelling through the Tunisian and Libyan desert. The impression must be particularly surreal from the southern Tunisian salt

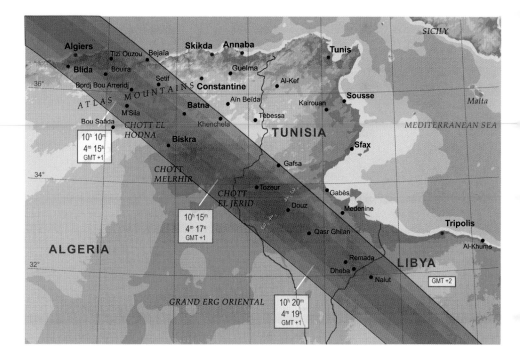

desert which in any case has something alien about it. The eclipse reaches its geographical midpoint in Sudan. Its speed has fallen to about 2 000 km/h. At the same time its duration there has increased to 4 m 21 s and the coverage to 92%. After traversing Sudan, its path turns eastwards through northern Kenya and ends in the Indian Ocean south of the Seychelles.

The Sun rises on the Portuguese coast at 07:30 (Western European Summer Time) and one hour later the eclipse begins. More and more of the Moon moves across the Sun, which rises higher and higher.

To the left, below the ring-shaped Sun, is a beautiful trio of luminaries: Jupiter, Mercury and Spica, the main star in the constellation of Virgo, are grouped together. Unfortunately, it will not be possible to observe this grouping since the remaining 8 or 10% of sunlight make the sky too bright.

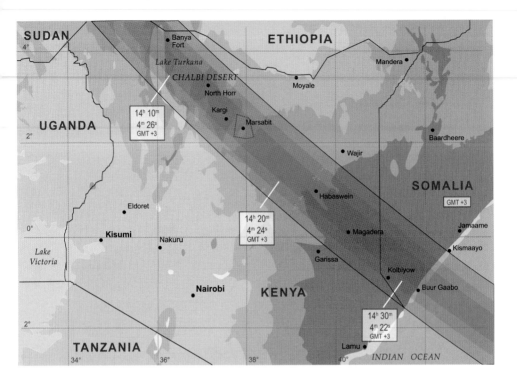

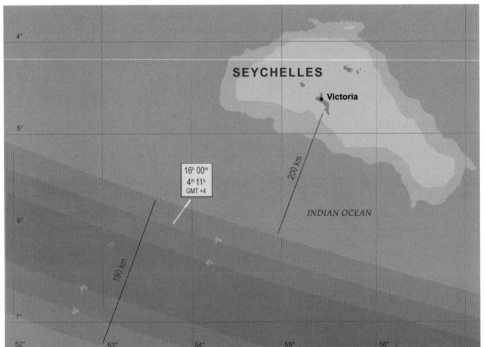

2005 Oct 3, Monday Location	Time zone (GMT ±)	Begin- ning	Maxi- mum	End	Dura- tion	Height of Sun	Cover- age
Sta Eugenia (Riberia), Spain	+2	09.38	10.53	12.15	4m 03s	23°	90%
Vigo, Spain	+2	09.38	10.53	12.16	4m 06s	24°	90%
Santiago de Comp., Spain	+2	09.39	10.53	12.16	2m 59s	23°	90%
Orense, Spain	+2	09.39	10.54	12.17	3m 50s	24°	90%
Braga, Portugal	+1	08.38	09.54	11.17	2m 54s	24°	90%
Bragança, Portugal	+1	08.39	09.55	11.19	4m 01s	25°	90%
Zamora, Spain	+2	09.39	10.56	12.21	4m 02s	26°	90%
Salamanca, Spain	+2	09.39	10.56	12.19	4m 02s	26°	90%
Valladolid, Spain	+2	09.40	10.56	12.22	1m 51s	27°	90%
Avila, Spain	+2	09.40	10.57	12.22	4m 06s	27°	90%
Segovia, Spain	+2	09.40	10.57	12.22	4m 09s	27°	90%
Madrid, Spain	+2	09.39	10.58	12.22	4m 09s	28°	90%
Toledo, Spain	+2	09.40	10.58	12.23	2m 43s	32°	91%
Carrascosa d Campo, Spain	+2	09.40	10.58	12.24	3m 55s	28°	90%
Cuenca, Spain	+2	09.41	10.58	12.24	3m 39s	32°	91%
Albacete, Spain	+2	09.41	11.01	12.28	3m 20s	32°	91%
Valencia, Spain	+2	09.42	11.02	12.30	3m 27s	32°	91%
Gandia, Spain	+2	09.43	11.02	12.31	3m 51s	32°	91%
Alicante, Spain	+2	09.42	11.02	12.30	3m 16s	32°	91%
Eivissa, Ibiza, Spain	+2	09.43	11.04	12.32	0m 57s	32°	91%
Algiers, Algeria	+1	08.45	10.07	11.38	3m 58s	36°	91%
Biskra, Algeria	+1	08.48	10.12	11.44	3m 05s	36°	91%
Batna, Algeria	+1	08.48	10.12	11.44	4m 02s	36°	91%
Gafsa, Tunisia	+1	08.50	10.16	11.50	1m 49s	36°	91%
Tozeur, Tunisia	+1	08.51	10.18	11.52	2m 27s	43°	91%
Douz, Algeria	+1	08.51	10.18	11.52	4m 15s	43°	91%
Remada, Algeria	+1	08.53	10.21	11.56	4m 19s	43°	91%
Nalut, Lybia	+1	08.54	10.22	11.58	4m 19s	43°	91%
Banya Fort, Kenya	+3	12.23	14.08	15.45	3m 59s	63°	92%
North Horr, Kenya	+3	12.29	14.14	15.49	3m 06s	61°	92%
Marsabit, Kenya	+3	12.32	14.16	15.51	4m 15s	60°	92%
Habaswein, Kenya	+3	12.38	14.22	15.56	4m 25s	57°	92%
Magadera, Kenya	+3	12.43	14.26	15.58	4m 24s	55°	92%
Buur Gaabo, Somalia	+3	12.49	14.31	16.02	4m 05s	52°	91%
300 km SSW of Seychelles	+4	14.31	16.03	17.21	4m 08s	32°	91%

The weather

In Spain, the probability of cloud cover increases during September. In October it is 30–50%, with conditions west of Salamanca being slightly less favourable. There, particularly near the Atlantic, 60% probability of cloud cover can be expected. At this time of year Galicia is one of the regions with the most rain in the whole of Europe. In the early autumn, however, the cloud cover is mostly broken, so clear patches can be expected even when there is bad weather. On Spain's eastern coast, there is a 35% probability of cloud, so there is a good chance that the sky will be clear or only covered by isolated clouds.

In North Africa, south of the Atlas Mountains, the cloud probability falls to about 30%. A grandiose view of the eclipse will undoubtedly be available from the Chott el Jerid, the large salt lake with its absolutely flat surface.

The best weather within the band of the eclipse is likely to be in northern Kenya, where there is only 20% probability of cloud. The conditions may be slightly better on the coast. An exotic location for observation of the eclipse is the Seychelles. There 86% of the Sun will be covered. A small, enchanting sickle of the Sun will be visible in the western sky there. In October, the rainy season has not yet started on the islands, so clear skies can be expected.

2006 March 29, total

Total: Brazil, West Africa, Libya, Egypt, Turkey, Georgia, Russia, Kazakhstan, Mongolia
Partial: Europe, Africa (except south) and south-west Asia

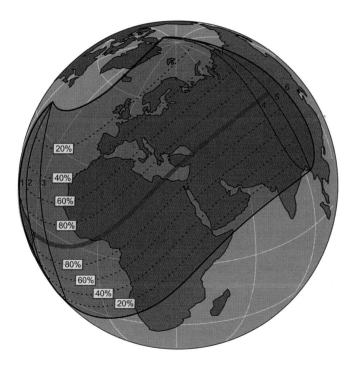

Course of the eclipse

1 End of eclipse at sunrise

2 Maximum at sunrise

3 Beginning of eclipse at sunrise

4 End of eclipse at sunset

5 Maximum at sunset

6 Beginning of eclipse at sunset

This eclipse, with a duration of almost 4 minutes, lasts nearly twice as long as the previous European total eclipse in 1999. Apart from the eclipse of 2015 visible in the Faeroes and Spitsbergen, this will be the total eclipse closest to Europe until 2026.

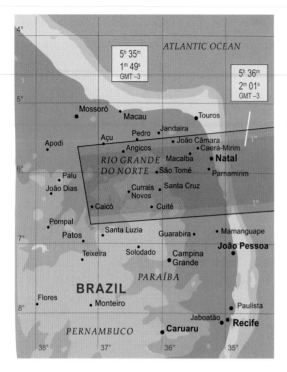

This eclipse starts at the eastern tip of Brazil, then travels across the Atlantic and carries on through the West African coastal states of Ghana, Togo, Benin and Nigeria, crossing the Sahara in Niger and Libya. The shadow leaves the African continent at the border between Libya and Egypt. Here the eclipse still has a duration of 4 minutes.

It travels across the Mediterranean between Crete and Cyprus, touching the mainland near the Turkish city of Antalya. Here the eclipse has a duration of 3 minutes 26 seconds with a totality zone of 165 km. Then the eclipse travels across the city of

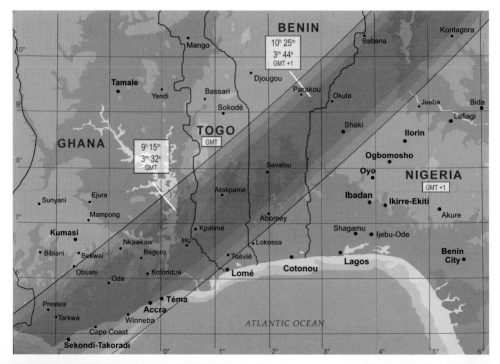

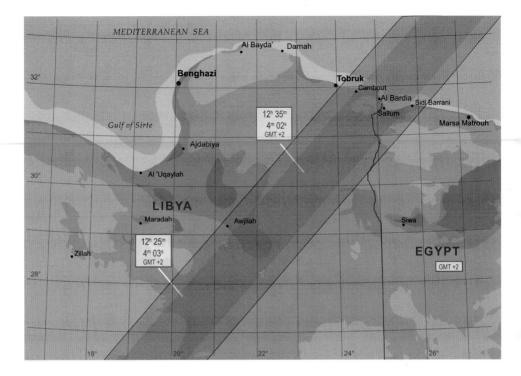

2006 March 29, Wednesday Location	Time zone (GMT ±)	Begin- ning	Maxi- mum	End	Dura- tion	Height of Sun
Caicó, Brazil	-3	05.35	05.35	09.33	0ᵐ 51ˢ	0°
Natal, Brazil	-3	05.27	05.36	06.35	1ᵐ 42ˢ	2°
Sekondi-Tak., Ghana	0	07.59	09.09	10.26	3ᵐ 23ˢ	44°
Accra, Ghana	0	08.01	09.12	10.30	2ᵐ 47ˢ	46°
Begoro, Ghana	0	08.02	09.13	10.31	3ᵐ 24ˢ	46°
Ho, Ghana	0	08.03	09.14	10.33	3ᵐ 00ˢ	46°
Kpalimé, Togo	0	08.04	09.15	10.34	3ᵐ 34ˢ	46°
Atakpamé, Togo	0	08.05	09.17	10.36	3ᵐ 30ˢ	49°
Savalou, Benin	+1	09.06	10.19	11.38	3ᵐ 37ˢ	50°
Shaki, Nigeria	+1	09.09	10.22	11.42	3ᵐ 11ˢ	53°
Gasau, Nigeria	+1	09.18	10.33	11.58	3ᵐ 51ˢ	58°
Katsina, Nigeria	+1	09.20	10.36	11.58	3ᵐ 51ˢ	59°
Maradi, Niger	+1	09.21	10.37	11.58	3ᵐ 11ˢ	59°
Zinder, Niger	+1	09.23	10.40	12.02	2ᵐ 43ˢ	60°
Tanout, Niger	+1	09.25	10.42	12.04	3ᵐ 44ˢ	61°
Bilma, Niger	+1	09.38	10.56	12.19	4ᵐ 03ˢ	65°
Awjilah, Libya	+2	11.10	12.30	13.51	2ᵐ 22ˢ	64°

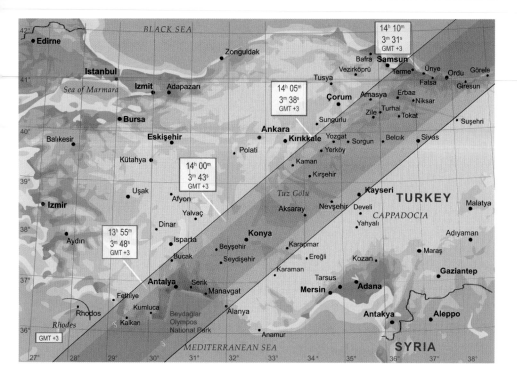

Map 1 (Turkey):

BLACK SEA

Edirne

Zonguldak

14ʰ 10ᵐ
3ᵐ 31ˢ
GMT +3

Bafra Samsun
Istanbul   Vezirköprü   Terme Ünye Ordu Görele
                          Tusya   Fatsa
Izmit Adapazarı   14ʰ 05ᵐ   Çorum Amasya Erbaa   Giresun
Sea of Marmara   3ᵐ 38ˢ   Niksar
                    GMT +3   Zile Turhal Tokat
Bursa   Sungurlu   Suşehri

Balıkesir   Eskişehir   Ankara
                          Kırıkkale Yozgat Sorgun Belcik Sivas
                    Polatı   Yerköy
Kütahya   Kaman
14ʰ 00ᵐ   Kırşehir
3ᵐ 43ˢ
GMT +3   Tuz Gölü   Kayseri TURKEY
Uşak   Nevşehir Develi   Malatya
Izmir   Afyon   Aksaray   CAPPADOCIA
                Yalvaç   Yahyalı
            Dinar   Adıyaman
Aydın   13ʰ 55ᵐ   Isparta   Konya
         3ᵐ 48ˢ   Beyşehir Karapınar   Maraş
         GMT +3   Bucak   Seydişehir   Ereğli Kozan
                              Karaman   Gaziantep
                Antalya Serik   Tarsus
                    Manavgat   Mersin Adana
Fethiye   Antakya   Aleppo
Rhodos   Kumluca Beydağlar   Alanya
Rhodes   Kalkan Olympos
GMT +3   National Park   Anamur   SYRIA
                MEDITERRANEAN SEA

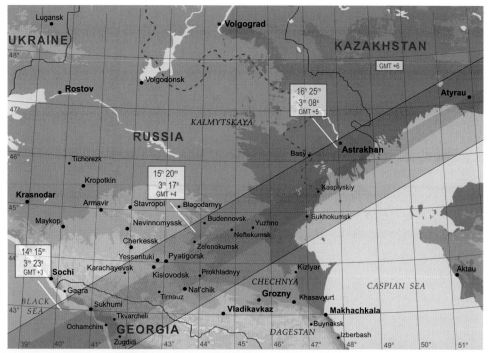

Map 2 (Russia/Caucasus):

Lugansk   Volgograd
UKRAINE   KAZAKHSTAN
                              GMT +6
Rostov   Atyrau
            Volgodonsk
                    KALMYTSKAYA   16ʰ 25ᵐ
RUSSIA   3ᵐ 08ˢ
                    Basy   GMT +5
Tichorezk   Astrakhan
    Kropotkin   15ʰ 20ᵐ   Kaspiyskiy
Krasnodar   3ᵐ 17ˢ
    Armavir   GMT +4
Maykop   Stavropol Blagodarnyy Sukhokumsk
        Nevinnomyssk Budennovsk
                Yuzhno
        Cherkessk   Neftekumsk
14ʰ 15ᵐ   Yessentuki Zelenokumsk
3ᵐ 23ˢ   Karachayevsk Pyatigorsk
GMT +3 Sochi   Kizlyar   Aktau
    Gagra   Kislovodsk Prokhladnyy
BLACK   Nal'chik CHECHNYA CASPIAN SEA
SEA   Sukhumi Tirnauz Grozny Khasavyurt
        Tkvarcheli   Vladikavkaz Makhachkala
Ochamchire GEORGIA   Buynaksk
        Zugdidi   DAGESTAN   Izberbash

2006 March 29, Wednesday Location	Time zone (GMT ±)	Begin-ning	Maxi-mum	End	Dura-tion	Height of Sun
Al Bardia, Libya	+2	11.20	12.40	14.00	3m 58s	62°
Sallum, Egypt	+2	11.20	12.40	14.00	3m 45s	62°
Sidi Barrani, Egypt	+2	11.21	12.41	14.01	1m 49s	62°
Kumluca, Turkey	+3	12.36	13.55	15.12	3m 43s	55°
Antalya, Turkey	+3	12.38	13.56	15.13	3m 26s	54°
Manavgat, Turkey	+3	12.38	13.57	15.14	3m 41s	54°
Alanya, Turkey	+3	12.39	13.57	15.14	1m 52s	54°
Konya, Turkey	+3	12.42	13.57	15.14	3m 41s	53°
Aksaray, Turkey	+3	12.45	14.03	15.18	3m 23s	51°
Kirsehir, Turkey	+3	12.46	14.04	15.19	3m 28s	50°
Yozgat, Turkey	+3	12.48	14.05	15.20	2m 55s	50°
Amasya, Turkey	+3	12.50	14.07	15.21	3m 41s	48°
Sivas, Turkey	+3	12.51	14.08	15.22	1m 36s	48°
Tokat, Turkey	+3	12.51	14.08	15.23	3m 34s	48°
Ordu, Turkey	+3	12.54	14.10	15.24	3m 30s	46°
Sukhumi, Georgia	+3	13.00	14.16	15.27	3m 09s	43°
Zugdidi, Georgia	+3	13.01	14.16	15.28	3m 30s	43°
Nal'chik, Russia	+4	14.04	15.19	16.30	2m 56s	41°
Astrakhan, Russia	+5	15.12	16.25	17.33	1m 39s	36°
Atyrau, Kazakhstan	+6	16.17	17.29	18.36	2m 59s	33°
Sagiz, Kazakhstan	+6	16.21	17.31	18.37	2m 30s	30°
Astana/Karaghandy, Kaz'stan	+7	17.37	18.42	19.42	2m 19s	18°
Rubtovsk, Russia	+8	18.43	19.45	20.43	2m 00s	11°
Biysk, Russia	+7	17.46	18.46	19.42	2m 08s	9°
Gorno-Altaysk, Russia	+7	17.46	18.46	19.39	2m 07s	8°
Choya, Russia	+7	17.46	18.46	19.36	2m 06s	8°
Kubayka, Russia	+7	17.47	18.46	19.23	2m 06s	6°
Kyzyl, Russia	+7	17.49	18.47	19.04	1m 53s	3°
Kazantsevo, Russia	+7	17.50	18.47	19.00	1m 39s	2°

Konya and just touches Kayseri on the Anatolian plateau as well as the great salt lake, Tuz Gölü.

Once the eclipse has crossed the Black Sea, its path crosses the north Georgian coast, passes the city of Sukhumi and travels across the northern part of Chechnya. Then the core shadow touches the Caspian Sea, passing Astrakhan in Russia and Atyrau in Kazakhstan. Here the eclipse only lasts for 3 minutes even at the centre of the totality. Further east the eclipse is visible for less than 3 minutes. The solar eclipse comes to an end in southern Russia towards Mongolia.

The weather

Ghana and Benin, where the band of the eclipse reaches the African continent, have strong cloud cover in March with a probability of 70%. However, the rainy season has not yet started. The conditions for observation of the eclipse are undoubtedly best in in Libya, where the degree of cloud cover lies at 30% in the north and 20% in the south. The sparsity of flora and fauna in the desert makes this area less attractive, however, since the reaction of surrounding nature is part of the observation of the eclipse.

Away from the coast, eastern Turkey is subject to a continental climate with little rain. The time of the year with most rain ends in March. Central Turkey has an average of only three days of rain at this time of year. It is best to observe the eclipse from the Anatolian highlands or the south coast, since the chance of cloud cover increases in the north close to the Black Sea. In Georgia and Kazakhstan, the sky will be clouded over with a probability of 60 to 70% at the end of March, although little rain falls there.

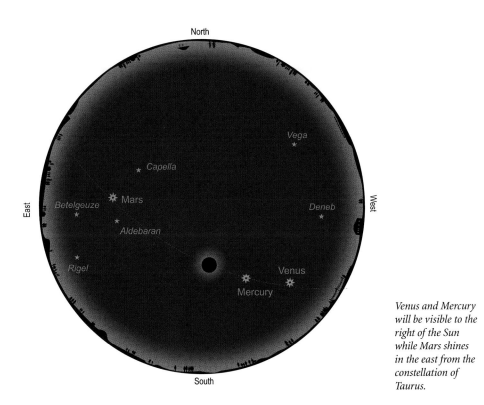

Venus and Mercury will be visible to the right of the Sun while Mars shines in the east from the constellation of Taurus.

2006 September 22, annular

Annular: Surinam, French Guiana, Brazil and South Atlantic
Partial: South America and western and southern Africa.

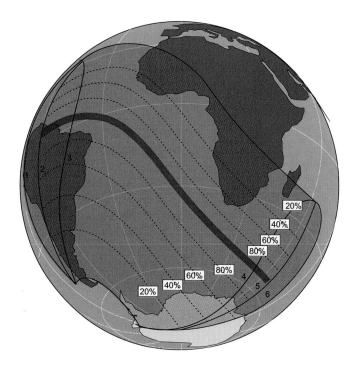

Course of the eclipse

1 *End of eclipse at sunrise*

2 *Maximum at sunrise*

3 *Beginning of eclipse at sunrise*

4 *End of eclipse at sunset*

5 *Maximum at sunset*

6 *Beginning of eclipse at sunset*

This annular solar eclipse takes place in the southern hemisphere at the start of spring. Since the apogee of the Moon is only 6 hours before the New Moon position, the covering Moon is extremely 'small,' therefore 12% of the Sun remains uncovered and a relatively wide ring of the Sun triumphs over the dark Moon.

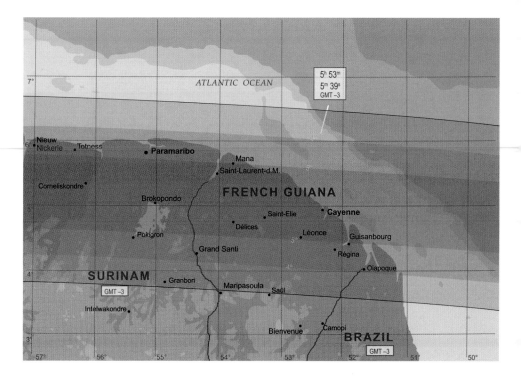

The central zone of this eclipse travels almost entirely over the ocean. Only at the start can it be seen for a very short time at sunrise on the South American coast of Surinam and French Guiana. An impressive view of the rise of the Sun's halo will be had particularly in places which have an unrestricted view of the ocean. As a partial eclipse, it can be seen from the whole of the South American continent except the most southern part. In Africa, it can be seen as partial in the west and southern Africa.

The 270 km wide band of the eclipse crosses the South Atlantic, passing 300 km to the south of Prince Edward Island and Crozet Islands. The annular solar eclipse ends shortly before Heart Island in the southern Indian Ocean.

About 12% of the Sun remains uncovered in this annular solar eclipse, so the planets will probably not be visible even if it is observed from the central zone in the Atlantic. All the classic planets are gathered around the Sun. To the east (below), Mars and Mercury are positioned close to the Sun with Venus to the

2006 Sep 22, Friday Location	Time zone (GMT ±)	Beginning	Maximum	End	Duration	Height of Sun	Coverage
Comeliskondre, Surinam	−3	06.37	06.52	08.07	5m 29s	4°	85%
Paramaribo, Surinam	−3	06.33	06.52	08.07	4m 55s	4°	85%
Brokopondo, Surinam	−3	06.33	06.52	08.08	5m 32s	5°	85%
St Laurent, French Guiana	−3	06.29	06.52	08.08	5m 24s	5	85%
Délices, French Guiana	−3	06.28	06.52	08.09	5m 24s	6°	85%
Léonce, French Guiana	−3	06.24	06.52	08.11	5m 14s	7°	85%
Cayenne, French Guiana	−3	06.22	06.53	08.11	5m 38s	8°	85%
Oiapoque, Amapá, Brazil	−3	06.20	06.53	08.12	3m 27s	9°	85%

west (above). At a greater distance we have Jupiter and Saturn, forming a symmetrical frame.

The weather

The conditions for observation of the eclipse are quite favourable because September is the month with least rainfall on the north coast of South America. The sky will most likely be covered by a few single clouds, but this is unlikely to be detrimental to observation of the eclipse.

2008 February 7, annular

Annular: Antarctic,
Partial: South-eastern Australia, New Zealand, Antarctica,
  southern Polynesia

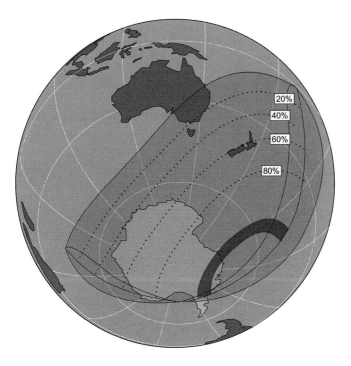

Course of the eclipse

1 End of eclipse at sunrise

2 Maximum at sunrise

3 Beginning of eclipse at sunrise

4 End of eclipse at sunset

5 Maximum at sunset

6 Beginning of eclipse at sunset

The eclipse only appears annular when seen from the Antarctic. The Sun is covered to 96.6%, so that for about 2 minutes and 12 seconds only a small ring of light is left to shine over the icy

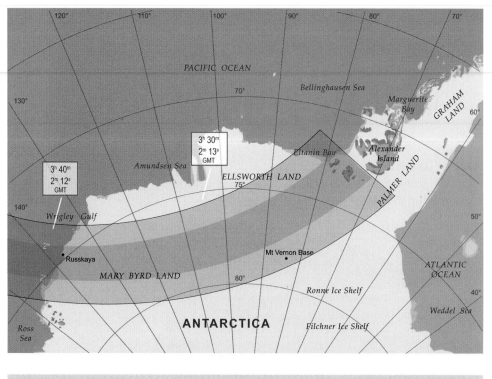

2008 Feb 7, Thursday	Time zone	Begin-	Maxi-	End	Dura-	Height	Cover-
Location	(GMT ±)	ning	mum		tion	of Sun	age
Russkaya, Antarctica	0	02.39	03.40	04.40	2ᵐ 11ˢ	14°	93%
Mt Vernon Base, Antarctica	0	02.26	03.26	04.21	1ᵐ 30ˢ	6°	93%

landscape. In the polar regions of the Earth the cone of the eclipse is flattened, so the path of the eclipse is correspondingly long. In this eclipse the annular shadow can be seen in a band which is about 447 km wide. This will only be of relevance to a very few committed observers, however, as reaching this site most definitely takes on the character of an expedition. The eclipse appears as partial in southern Australia, New Zealand and southern Polynesia.

Saros series 121, of which the current eclipse is a part, now approaches its end. Two representative of this series will still occur as annular solar eclipses. Thereafter the lunar shadow

SOLAR ECLIPSES

travels so far from the Earth that only a part of the penumbra still touches Antarctica. This eclipse series ends in 2206 after 1200 years.

The conjunction of Mercury with the Sun almost coincides with the timing of the eclipse. As a consequence this closest planet to the Sun remains invisible. Venus and Jupiter, not far to the left in the constellation of Sagittarius, will probably be able to assert themselves against the brightness of the remaining corona of the Sun.

2008 August 1, total

Total: North-east Canada, northern Greenland, central Siberia, Mongolia, northern China
Partial: North-east Canada, Europe (except south-west), Asia

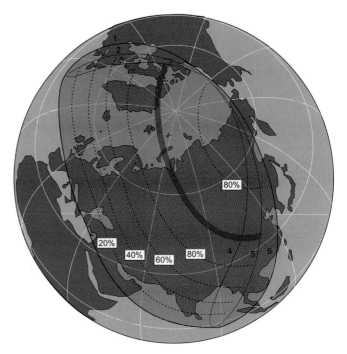

Course of the eclipse

1 End of eclipse at sunrise

2 Maximum at sunrise

3 Beginning of eclipse at sunrise

4 End of eclipse at sunset

5 Maximum at sunset

6 Beginning of eclipse at sunset

The eclipse begins in the eternal ice of north-eastern Canada and then touches northern Greenland. Then its path travels between Spitsbergen (Svalbard) and the islands of Franz Joseph Land, crosses the sickle-shaped island of Novaya Zemlya and

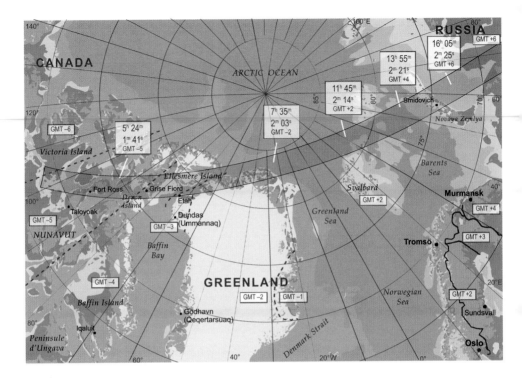

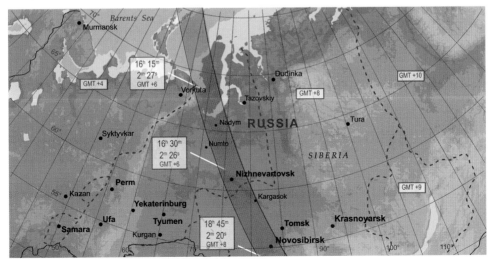

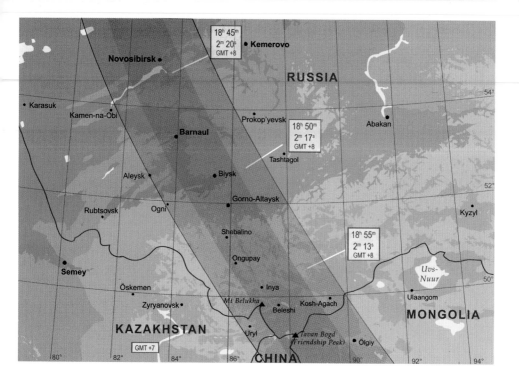

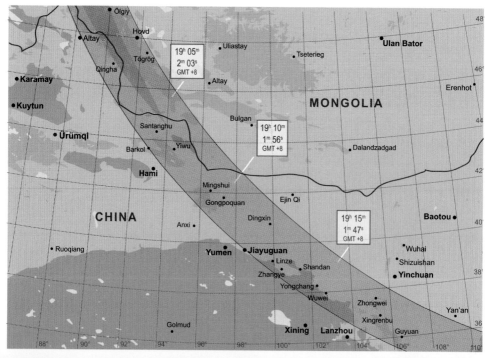

2008 Aug 1, Friday Location	Time zone (GMT ±)	Begin-ning	Maxi-mum	End	Dura-tion	Height of Sun
Fort Ross, Canada	−5	03.31	04.23	05.16	1ᵐ 26ˢ	5°
Grise Fjord, Canada	−4	04.31	05.25	06.20	1ᵐ 38ˢ	10°
Etah, Greenland	−3	05.31	06.27	07.22	0ᵐ 50ˢ	13°
Smidovich, Novaya Zemlya	+4	12.57	14.00	15.02	2ᵐ 14ˢ	31°
Nadym, Russia	+6	15.17	16.21	17.23	2ᵐ 27ˢ	34°
Nizhnevartovsk, Russia	+6	15.27	16.31	17.33	2ᵐ 19ˢ	33°
Novosibirsk, Russia	+8	17.41	18.45	19.45	2ᵐ 19ˢ	30°
Barnaul, Russia	+8	17.45	18.48	19.48	2ᵐ 14ˢ	30°
Biysk, Russia	+8	17.48	18.51	19.51	2ᵐ 14ˢ	28°
Gorno-Altaysk, Russia	+8	17.49	18.52	19.41	2ᵐ 13ˢ	30°
Onguday, Russia	+8	17.51	18.54	19.53	2ᵐ 08ˢ	28°
Inya, Russia	+8	17.52	18.55	19.54	2ᵐ 11ˢ	27°
Uryl, Kazakhstan	+7	16.54	17.56	18.56	0ᵐ 43ˢ	27°
Beleshi, Russia	+8	17.54	18.56	19.55	2ᵐ 10ˢ	26°
Kosh Agach, Russia	+8	17.54	18.57	19.54	1ᵐ 45ˢ	27°
Ölgiy, Mongolia	+8	17.57	18.59	19.56	1ᵐ 39ˢ	24°
Altay, Xinjiang, China	+8	17.58	18.56	19.58	1ᵐ 20ˢ	24°
Hovd, Mongolia	+8	18.01	19.02	20.00	2ᵐ 04ˢ	23°
Gongpoquan, Gansu, China	+8	18.13	19.11	20.06	1ᵐ 53ˢ	17°
Zhangye, Gansu, China	+8	18.19	19.15	20.09	1ᵐ 26ˢ	13°
Luoyang, Henan, China	+8	18.26	19.20	19.28	1ᵐ 06ˢ	2°

leaves the Arctic Ocean towards Novosibirsk in Siberia and northern China. There the solar eclipse ends. At 2 minutes 27 seconds, this event is one of the shorter eclipses. The date makes it very likely that there will be good visibility in the Arctic Ocean and Greenland.

The next representative of the Saros series, of which this eclipse is a part, will be the one occurring in 2026, which will pass through northern Spain as well as the favourite holiday destination of Majorca. Since this eclipse will also happen in July, huge crowds can be expected.

As soon as the Moon has fully covered the Sun, it is worth looking slightly to the left of the Sun. At a distance of about 6°, Mercury, Venus, Regulus (the main star in Leo), and then Saturn and Mars can be seen in a line. This uniquely highlights the constellation of Leo in the zodiac.

Weather

In northern Canada as well as in Greenland, there is a 70% probability that the sky will be covered with cloud in August. In Nova Zemlya the conditions are even worse, with an average of more than 20 days of rainfall in August. South of Novosibirsk, through which the central line passes, the possibilities for observing the eclipse improve significantly. The cloud banks drop their rain in the Mongolian mountain ranges and thus lead to cover of only about 30% south of latitude 45°, in Mongolia.

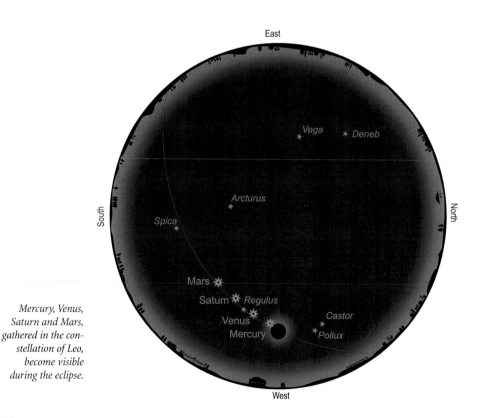

Mercury, Venus, Saturn and Mars, gathered in the constellation of Leo, become visible during the eclipse.

2009 January 26, annular

Annular: Indonesia
Partial: Southern Africa, southern and eastern India, Indo-China, Australia (except south-west and Tasmania), Antarctica

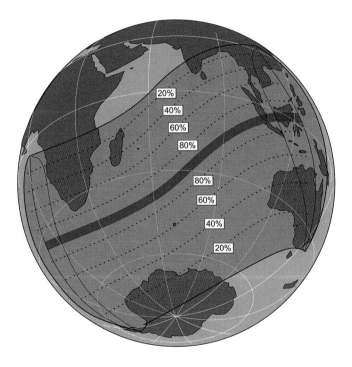

Course of the eclipse

1 End of eclipse at sunrise

2 Maximum at sunrise

3 Beginning of eclipse at sunrise

4 End of eclipse at sunset

5 Maximum at sunset

6 Beginning of eclipse at sunset

The eclipse starts to the south of Africa, travels through the southern Indian Ocean and only reaches land at its end. It passes directly through the Sunda Straits between Sumatra and Java, including the mighty Krakatau volcano. Jakarta lies on the edge

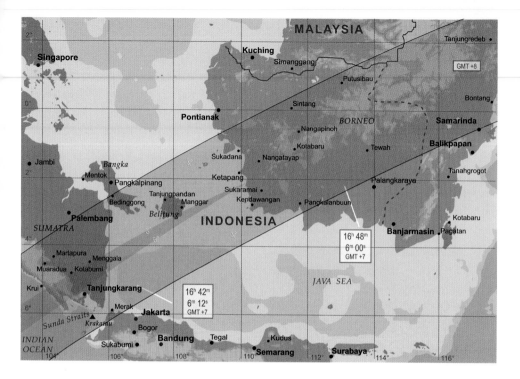

of the shadow zone. Then the island of Belitung is traversed before the path of the eclipse crosses Borneo. At its end, the eclipse reaches a speed of 10 000 km/h. Over Borneo only 85% of the Sun's disk is covered. Such an astronomical event could certainly provide an occasion to take a trip to the Far East to marvel at the wonder of an annular solar eclipse in which a ring of sunlight remains around the dark grey New Moon, thus impressively demonstrating the victory of light.

The weather

The eclipse occurs in January and thus in the middle of the rainy season lasting from November to March during which it rains on average for 3 weeks in each month. Rainfalls can extend for days in this tropical climate. In January, five times as much rain falls in Indonesia than in Europe. The humidity is also pretty high outside the rainy season. On Sumatra, the main rainy season takes place a little earlier from October to January, so that

2009 Jan 26, Monday Location	Time zone (GMT ±)	Beginning	Maximum	End	Duration	Height of Sun	Coverage
Krui, Sumatra, Indonesia	+7	15.19	16.41	17.52	5ᵐ 57ˢ	24°	85%
Tanjungkarang, Sumatra	+7	15.20	16.42	17.51	6ᵐ 02ˢ	24°	85%
Martapura, Sumatra, Indonesia	+7	15.21	16.42	17.53	6ᵐ 02ˢ	23°	85%
Menggala, Lanpung, Indonesia	+7	15.20	16.42	17.51	6ᵐ 02ˢ	23°	85%
Bedinggang, S-Bangka, Indon'ia	+7	15.25	16.45	17.54	4ᵐ 00ˢ	21°	85%
Tanjungpandan, Behtung	+7	15.26	16.45	17.54	5ᵐ 17ˢ	19°	85%
Manggar, Belitung, Indonesia	+7	15.26	16.46	17.54	5ᵐ 57ˢ	18°	85%
Kendawangan, W-Borneo	+7	15.29	16.47	17.53	5ᵐ 56ˢ	17°	85%
Pangkalanbuun, W-Borneo	+7	15.30	16.47	17.50	4ᵐ 15ˢ	15°	84%
Nangapinoh, W-Borneo, Indon'ia	+7	15.33	16.48	17.46	4ᵐ 34ˢ	13°	84%
Sintang, W-Borneo, Indonesia	+7	15.34	16.48	17.47	2ᵐ 05ˢ	14°	84%
Tewah, C-Borneo, Indonesia	+7	15.34	16.49	17.39	5ᵐ 36ˢ	12°	84%
Samarinda, E-Borneo, Indonesia	+8	16.37	17.50	18.25	2ᵐ 27ˢ	9°	84%
Sangkulirung, E-Borneo	+8	16.39	16.49	18.19	5ᵐ 39ˢ	7°	84%

conditions may be a little better. Incidentally, the eclipse occurs two days after the conjunction between Jupiter and the Sun. The brightly shining planet Jupiter may be seen directly to the west of (below) the Sun.

2009 July 22, total

Total: India, Nepal, Bangladesh, Bhutan, China, southernmost islands of Japan, western Pacific Ocean
Partial: Central, southern and eastern Asia, Polynesia

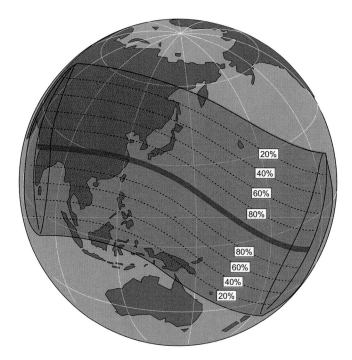

Course of the eclipse

1 End of eclipse at sunrise

2 Maximum at sunrise

3 Beginning of eclipse at sunrise

4 End of eclipse at sunset

5 Maximum at sunset

6 Beginning of eclipse at sunset

With a maximum duration of 6 minutes 39 seconds, this eclipse is among the longer ones. This is because on July 21 the Sun is very far from the Earth, therefore appearing smaller in the sky than in winter, while at the same time — and this is the key fac-

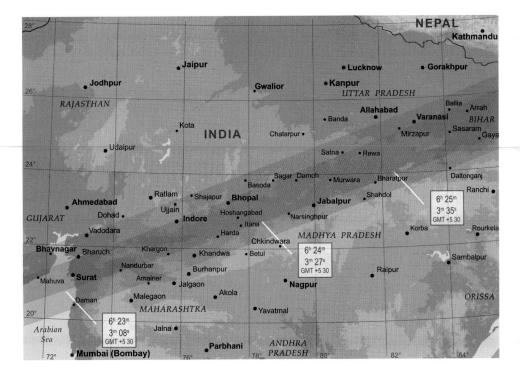

2009 July 22, Wednesday Location	Time zone (GMT ±)	Begin- ning	Maxi- mum	End	Dura- tion	Height of Sun
Bhavnagar, Gujarat, India	+5.30	06.14	06.23	07.20	1ᵐ 51ˢ	3°
Surat, Gujarat, India	+5.30	06.12	06.23	07.20	3ᵐ 05ˢ	3°
Bharuch, Gujarat, India	+5.30	06.11	06.23	07.20	2ᵐ 52ˢ	3°
Nandurbar, Maharashtra, India	+5.30	06.06	06.23	07.20	3ᵐ 00ˢ	4°
Khargon, Madhya Pradesh, India	+5.30	06.00	06.23	07.21	2ᵐ 59ˢ	5°
Indore, Madhya Pradesh, India	+5.30	05.58	06.23	07.22	3ᵐ 02ˢ	6°
Ujjain, Madhya Pradesh, India	+5.30	05.57	06.23	07.22	1ᵐ 29ˢ	6°
Khandwa, Madhya P. India	+5.30	05.57	06.23	07.22	2ᵐ 28ˢ	6°
Harda, Madhya Pradesh, India	+5.30	05.41	06.23	07.22	3ᵐ 05ˢ	7°
Bhopal, Madhya Pradesh, India	+5.30	05.50	06.24	07.23	3ᵐ 07ˢ	9°
Hoshangabad, Madhya P. India	+5.30	05.50	06.24	07.23	3ᵐ 19ˢ	7°
Basoda, Madhya Pradesh, India	+5.30	05.47	06.24	07.23	2ᵐ 06ˢ	9°
Narsinghpur, Madhya P. India	+5.30	05.44	06.24	07.24	2ᵐ 56ˢ	9°
Damoh, Madhya Pradesh, India	+5.30	05.41	06.24	07.23	3ᵐ 22ˢ	9°

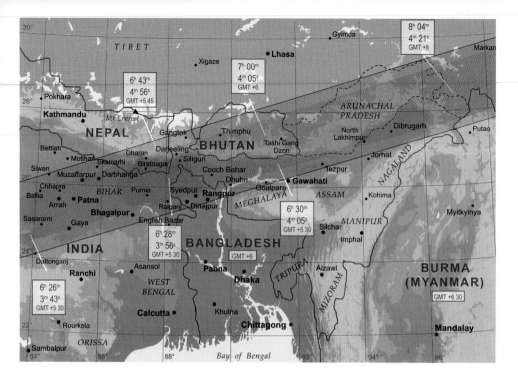

tor — the Moon is at perigee (closest to the Earth) on July 21, so that the Moon at 32' 22" is significantly larger than the Sun at 31' 29". These favourable size ratios mean that the path of the eclipse reaches the great width of 257 km. The eclipse begins on the west coast of India, passes across the Vindhya mountains of central India, touching south-east Nepal and northern Bangladesh before traversing the tiny Himalayan state of Bhutan and its capital Thimphu. The length of the eclipse on the subcontinent is 3 minutes 40 seconds. At the same time the shadow is still travelling at a speed of almost 10 000 km/h. In China the total eclipse throws its shadow over the huge cities of Wanxian, Wuhan and Shanghai. The eclipse approaches its midpoint and the speed of the shadow slows down to about 3000 km/h. Several of the small islands off Japan between Kyushu and Okinawa, such as the Tokara group are now traversed before the eclipse crosses the Pacific where the Gilbert Islands are thrown into darkness. The eclipse ends to the north-east of Western Samoa.

2009 July 22, Wednesday Location	Time zone Begin- (GMT ±) ning		Maxi- mum	End	Dura- tion	Height of Sun
Jabalpur, Madhya Pradesh, India	+5.30	05.40	06.24	07.24	3ᵐ 21ˢ	10°
Murwara, Madhya P. India	+5.30	05.37	06.25	07.24	3ᵐ 32ˢ	10°
Satna, Madhya Pradesh, India	+5.30	05.34	06.25	07.26	2ᵐ 59ˢ	11°
Rewa, Madhya Pradesh, India	+5.30	05.28	06.25	07.26	3ᵐ 23ˢ	12°
Bharatpur, Madhya P. India	+5.30	05.32	06.25	07.26	2ᵐ 52ˢ	11°
Mirzapur, Uttar Pradesh, India	+5.30	05.22	06.25	07.27	3ᵐ 05ˢ	12°
Varanasi, Uttar Pradesh, India	+5.30	05.23	06.26	07.28	3ᵐ 01ˢ	13°
Sasaram, Bihar, India	+5.30	05.23	06.26	07.28	3ᵐ 41ˢ	14°
Ballia, Uttar Pradesh, India	+5.30	05.24	06.26	07.29	2ᵐ 48ˢ	15°
Arrah, Bihar, India	+5.30	05.24	06.26	07.28	3ᵐ 35ˢ	15°
Gaya, Bihar, India	+5.30	05.24	06.26	07.28	3ᵐ 06ˢ	15°
Patna, Bihar, India	+5.30	05.24	06.26	07.29	3ᵐ 45ˢ	15°
Bhagalpur, Bihar, India	+5.30	05.25	06.27	07.30	2ᵐ 46ˢ	17°
Purnia, Bihar, India	+5.30	05.30	06.28	07.31	3ᵐ 41ˢ	18°
Biratnagar, Nepal	+5.45	05.45	06.43	07.47	3ᵐ 26ˢ	18°
Dharan, Nepal	+5.45	05.45	06.44	07.47	2ᵐ 39ˢ	18°
Raiganj, India	+5.30	05.30	06.28	07.32	2ᵐ 55ˢ	18°
Dinajpur, Rajshahi, Bangladesh	+6	05.56	06.58	08.02	2ᵐ 23ˢ	18°
Darjeeling, West Bengal, India	+5.30	05.31	06.28	07.33	2ᵐ 55ˢ	19°
Siliguri, West Bengal, India	+5.30	05.31	06.28	07.33	3ᵐ 45ˢ	19°
Syedpur, Rajshahi, Bangladesh	+6	05.56	06.58	08.02	2ᵐ 45ˢ	19°
Rangpur, Rajshahi, Bangladesh	+6	05.57	06.59	08.03	2ᵐ 07ˢ	19°
Cooch Behar, West Bengal, India	+5.30	05.31	06.29	07.34	3ᵐ 45ˢ	20°
Dhuburi, Assam, India	+5.30	05.31	06.29	07.34	2ᵐ 30ˢ	20°
Thimphu, Bhutan	+6	06.01	06.59	08.05	2ᵐ 58ˢ	19°
Tashi Gang Dzon, Bhutan	+6	06.01	07.00	08.07	4ᵐ 06ˢ	22°
Tezpur, Assam, India	+5.30	05.31	06.31	07.38	1ᵐ 50ˢ	23°
North Lakhimpor, Assam, India	+5.30	05.31	06.33	07.41	2ᵐ 39ˢ	25°
Dibrugarh, Assam, India	+5.30	05.32	06.33	07.41	3ᵐ 32ˢ	25°
Markam, Tibet	+8	08.04	09.07	10.17	2ᵐ 59ˢ	30°
Litang, Sichuan, China	+8	08.04	09.07	10.17	3ᵐ 00ˢ	30°
Shimian, Sichuan, China	+8	08.05	09.10	10.23	4ᵐ 28ˢ	34°
Kangding, Sichuan, China	+8	08.06	09.10	10.23	4ᵐ 11ˢ	34°
Zigong, Sichuan, China	+8	08.06	09.13	10.27	4ᵐ 01ˢ	37°
Chengdu, Sichuan, China	+8	08.07	09.13	10.26	3ᵐ 20ˢ	37°

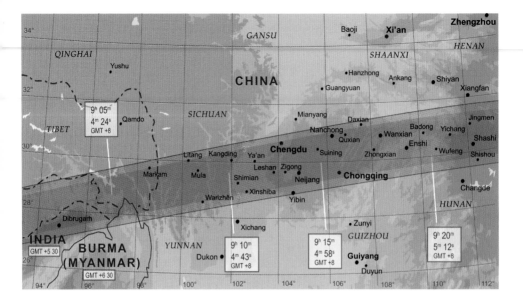

China, above all, is suitable for observing the eclipse because there the length of the shadow is sufficiently pronounced. The Gilbert Islands of Kiribati are also suitable.

The weather

In India, there is only about 10% probability of a clear sky at the height of summer. The conditions are better in western China, where the probability of a clear sky — or at least minimum cloud cover — lies at 25%.

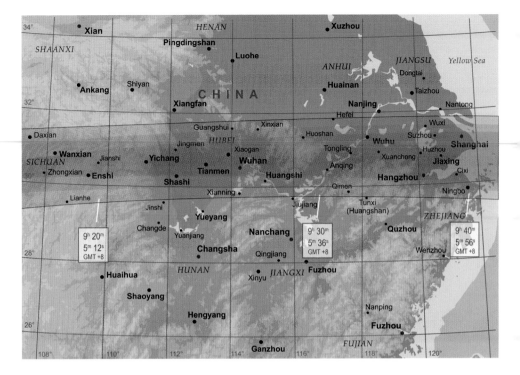

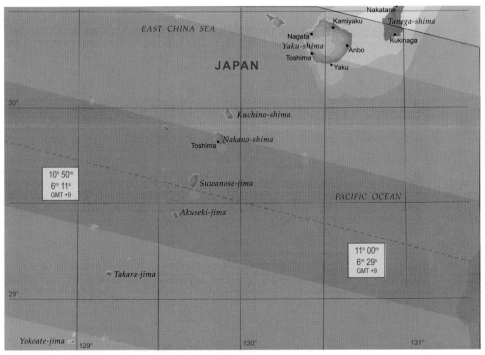

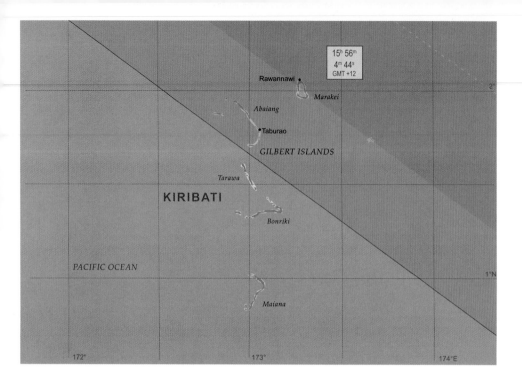

15ʰ 56ᵐ
4ᵐ 44ˢ
GMT +12

Rawannawi •

Marakei

Abaiang

• Taburao

GILBERT ISLANDS

Tarawa

KIRIBATI

Bonriki

PACIFIC OCEAN

Maiana

172°   173°   174°E

2009 July 22, Wednesday Location	Time zone (GMT ±)	Beginning	Maximum	End	Duration	Height of Sun
Neijang, Sichuan, China	+8	08.07	09.14	10.28	4m 28s	37°
Suining, Sichuan, China	+8	08.08	09.14	10.29	4m 29s	38°
Nanchong, Sichuan, China	+8	08.08	09.15	10.30	3m 54s	39°
Chongqing, Chongqing, China	+8	08.08	09.15	10.31	4m 04s	39°
Daxian, Sichuan, China	+8	08.10	09.17	10.33	2m 40s	40°
Zhongxian, Sichuan, China	+8	08.09	09.17	10.34	5m 04s	40°
Wanxian, Chonqing, China	+8	08.10	09.18	10.34	4m 42s	40°
Enshi, Hubei, China	+8	08.10	09.19	10.36	5m 03s	40°
Yichang, Hubei, China	+8	08.12	09.22	10.40	5m 16s	42°
Shashi, Hubei, China	+8	08.13	09.24	10.42	4m 56s	45°
Tianmen, Hubei, China	+8	08.14	09.25	10.44	5m 22s	46°
Xiaogan, Hubei, China	+8	08.15	09.26	10.45	5m 24s	47°
Guangshui, Hubei, China	+8	08.15	09.26	10.45	3m 37s	47°
Xianning, Henan, China	+8	08.15	09.27	10.46	2m 57s	47°
Wuhan, Hubei, China	+8	08.15	09.27	10.46	5m 23s	48°
Huangshi, Hubei, China	+8	08.15	09.28	10.48	5m 12s	48°
Huoshan, Anhui, China	+8	08.17	09.30	10.50	4m 49s	50°
Anqing, Anhui, China	+8	08.18	09.31	10.52	5m 23s	51°
Qimen, Anhui, China	+8	08.18	09.33	10.54	2m 49s	52°
Tongling, Anhui, China	+8	08.19	09.33	10.54	5m 38s	52°
Wuhu, Anhui, China	+8	08.20	09.34	10.55	5m 00s	52°
Hangzhou, Zhejiang, China	+8	08.21	09.37	10.59	5m 14s	55°
Suzhou, Jiangsu, China	+8	08.22	09.38	11.00	4m 53s	56°
Shanghai, China	+8	08.23	09.39	11.02	5m 01s	56°
Ningbo, Zhejiang, China	+8	08.23	09.40	11.03	4m 17s	56°
Toshima, Yaku-shima, Japan	+9	09.36	10.57	12.22	6m 01s	67°
Yaku, Yaku-shima, Japan	+9	09.37	10.58	12.23	4m 00s	68°
Kukinaga, Tanega-shima, Japan	+9	09.38	10.59	12.23	1m 19s	68°
Rawannawi, Gilbert Is., Kiribati	+12	14.44	15.57	17.02	4m 00s	37°

2010 January 15, annular

Annular: Central African Republic, Dem. Rep. of Congo, Uganda, Kenya, Maldives, Southern India, Burma, Sri Lanka, China
Partial: Africa (except west), South-East Europe, Asia (except far north-east)

Course of the eclipse

1 End of eclipse at sunrise

2 Maximum at sunrise

3 Beginning of eclipse at sunrise

4 End of eclipse at sunset

5 Maximum at sunset

6 Beginning of eclipse at sunset

This eclipse takes place just two days before the apogee of the Moon so that the Moon appears relatively small from the Earth. Its apparent diameter is 29' 15". In addition, the Sun is relatively close to the Earth because perigee is in the early days of January.

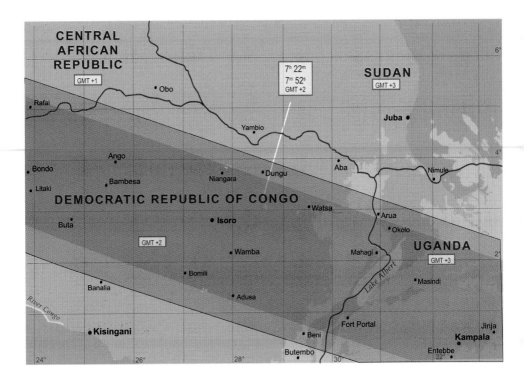

CENTRAL AFRICAN REPUBLIC
GMT +1

SUDAN
GMT +3

7ʰ 22ᵐ 7ᵐ 52ˢ GMT +2

Rafai
Obo
Yambio
Juba

Ango
Bondo
Bambesa
Niangara
Dungu
Aba
Nimule

Litaki
DEMOCRATIC REPUBLIC OF CONGO
Watsa
Arua

Isoro
Okolo
UGANDA
GMT +3

Buta
GMT +2
Wamba
Mahagi

Banalia
Bomili
Masindi

River Congo
Adusa
Lake Albert

Kisingani
Beni
Fort Portal
Jinja

Butembo
Entebbe
Kampala

24°
26°
28°
30°
32°

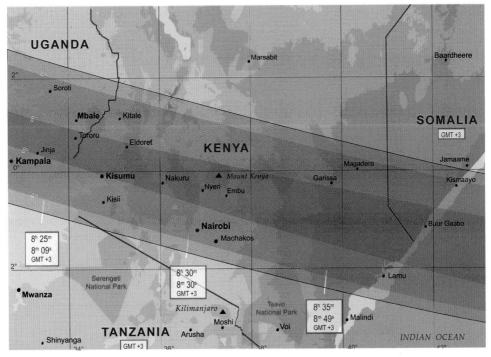

UGANDA
Marsabit
Baardheere

Soroti

Mbale
Kitale
SOMALIA
GMT +3

Tororo
Eldoret
KENYA

Jinja
Magadera
Jamaame

Kampala
Kisumu
Nakuru
Mount Kenya
Garissa
Kismaayo

Kisii
Nyeri
Embu

Buur Gaabo

8ʰ 25ᵐ 8ᵐ 09ˢ GMT +3

Nairobi
Machakos

Mwanza
Serengeti National Park
8ʰ 30ᵐ 8ᵐ 30ˢ GMT +3
Lamu

Kilimanjaro
Tsavo National Park
8ʰ 35ᵐ 8ᵐ 49ˢ GMT +3

TANZANIA
Arusha
Moshi
Voi
Malindi
INDIAN OCEAN

Shinyanga
GMT +3
34°
36°
38°
40°
42°

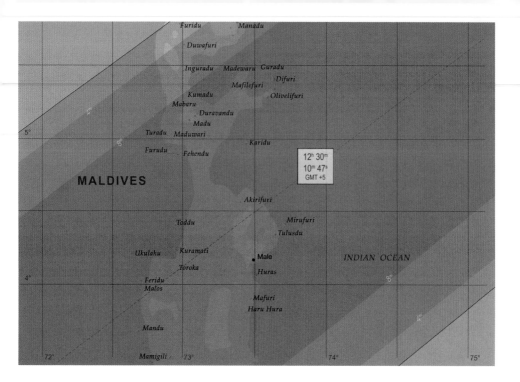

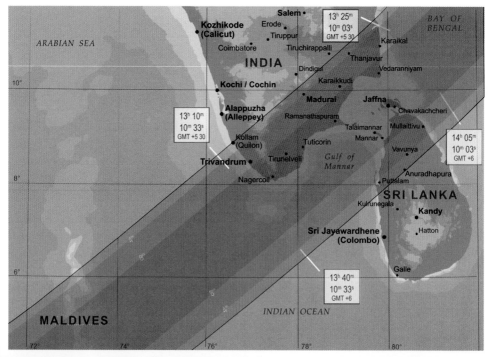

At 32' 31" the Sun is 11% larger than the Moon. The relative distance between Earth and Moon fluctuates four times as much as that between Earth and Sun. When both ratios occur at the same time (perigee of the Sun, apogee of the Moon) an annular solar eclipse occurs in which the ring of the Sun is particularly large. Cover is just 85%. At the same time these circumstances mean that the eclipse lasts for a particularly long time. In the middle of the eclipse — in the Indian Ocean — the ring of the Sun and New Moon can be seen for 11 minutes and 10 seconds. Such a long annular eclipse will not occur again for over a millennium until 3043.

This long eclipse begins in the western Central African Republic, moves across its capital Bangui, travels onward through the north of the Democratic Republic of Congo and crosses Lake Victoria in Uganda and its capital Kampala. Kenya and southern Somalia are the last African countries through which the eclipse travels. Here the cities of Kisumu, Nakuru and Nairobi as well as Mount Kenya are covered by the path of the shadow. At this point the eclipse already lasts $8^1/_2$ minutes.

Before the shadow zone moves across the southern tip of India and the north of Sri Lanka, the island group of the Maldives is also cast in shadow. At the maximum the Sun will have reached an elevation of 66° that is almost at its highest. At this time of the year, the probability of cloud on the Maldives is negligible.

Shortly afterwards the shadow moves across the southern tip of India and the northern part of Sri Lanka before crossing the Bay of Bengal and making a landfall in Myanmar (Burma). It travels across the city of Mandalay, and then on to China. After the total eclipse of July 2009, China is once again in the path of an eclipse and the rare event occurs that a city like Chongqing is able to enjoy two eclipses so close together. On average any place on Earth is traversed by a total eclipse only once every 300 years. The fact that south west China lies twice in the central path illustrates that such averages apply only over long periods of time. The eclipse ends on the Chinese peninsula of Shandong on the Yellow Sea.

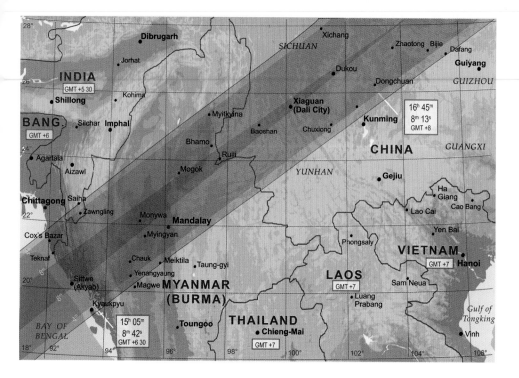

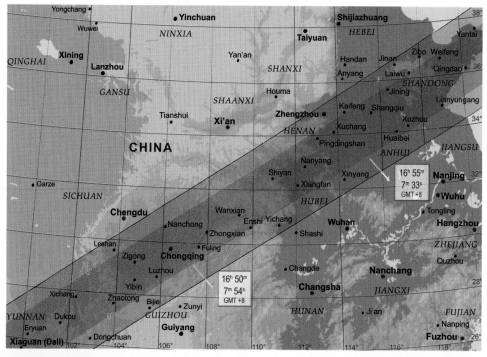

The weather

Where Kenya has a 90% chance of being cloudy, the sky is likely to be clear or at least only have slight cloud cover in Myanmar (Burma) with a cloud probability of 30%. The conditions deteriorate in China with 70% probability that the sky will have strong cloud cover.

2010 Jan 15, Friday Location	Time zone (GMT ±)	Begin- ning	Maxi- mum	End	Dura- tion	Height of Sun	Cover- age
Bangui, Central African Rep	+1	06.02	06.18	07.40	3ᵐ 55ˢ	3°	82%
Likati, Dem Rep Congo	+2	06.38	07.20	08.48	7ᵐ 16ˢ	10°	83%
Buta, Dem Rep Congo	+2	06.35	07.20	09.48	6ᵐ 52ˢ	10°	83%
Bomili, Dem Rep Congo	+2	06.24	07.21	08.58	5ᵐ 53ˢ	13°	83%
Fort Portal, Uganda	+3	07.09	08.24	09.58	6ᵐ 26ˢ	17°	83%
Masindi, Uganda	+3	07.06	08.25	10.08	7ᵐ 43ˢ	18°	83%
Entebbe, Uganda	+3	07.06	08.26	10.08	6ᵐ 52ˢ	20°	83%
Kampala, Uganda	+3	07.06	08.26	10.08	7ᵐ 43ˢ	20°	83%
Jinja, Uganda	+3	07.06	08.26	10.08	8ᵐ 07ˢ	20°	83%
Mbale, Uganda	+3	07.06	08.27	10.09	7ᵐ 38ˢ	21°	83%
Kisumu, Kenya	+3	07.06	08.28	10.19	8ᵐ 07ˢ	23°	83%
Nakaru, Kenya	+3	07.07	08.29	10.19	8ᵐ 25ˢ	23°	83%
Nairobi, Kenya	+3	07.07	08.29	10.14	6ᵐ 45ˢ	23°	83%
Garissa, Kenya	+3	07.08	08.34	10.24	8ᵐ 09ˢ	24°	83%
Magadera, Kenya	+3	07.09	08.35	10.24	5ᵐ 42ˢ	25°	83%
Lamu, Kenya	+3	07.08	08.37	10.24	7ᵐ 12ˢ	29°	83%
Buur Gaabo, Somalia	+3	07.09	08.38	10.32	8ᵐ 48ˢ	31°	84%
Kismaayo, Somalia	+3	04.10	08.39	10.32	4ᵐ 55ˢ	33°	84%
Male, Maldives	+5	10.15	12.25	14.23	10ᵐ 30ˢ	65°	85%
Trivandrum, Kerala, India	+5.30	11.05	13.15.	15.06.	7ᵐ 12ˢ	58°	84%
Nagercoil, Tamil Nadu, India	+5.30	11.06	13.16	15.06	9ᵐ 51ˢ	58°	84%
Tirunelveli, Tamil Nadu, India	+5.30	10.08	13.17	15.07	8ᵐ 57ˢ	58°	84%
Tuticorin, Tamil Nadu, India	+5.30	11.10	13.18	15.08	9ᵐ 45ˢ	58°	84%
Karaikkudi, Tamil Nadu, India	+5.30	11.15	13.23	15.10	7ᵐ 37ˢ	56°	84%
Thanjavur, Tamil Nadu, India	+5.30	11.17	13.25	15.12	3ᵐ 27ˢ	55°	84%
Vedaranniyam, Tamil N. India	+5.30	11.19	13.26	15.12	9ᵐ 16ˢ	55°	84%
Karaikal, Tamil Nadu, India	+5.30	11.20	13.27	15.12	7ᵐ 13ˢ	54°	84%

2010 Jan 15, Friday Location	Time zone (GMT ±)	Begin- ning	Maxi- mum	End	Dura- tion	Height of Sun	Cover- age
Puttalam, Sri Lanka	+6	11.46	13.52	15.39	7ᵐ 15ˢ	57°	84%
Mannar, Sri Lanka	+6	11.46	13.53	15.40	9ᵐ 27ˢ	56°	84%
Vavunya, Sri Lanka	+6	11.48	13.54	15.41	6ᵐ 19ˢ	56°	84%
Jaffna, Sri Lanka	+6	11.48	13.55	15.42	10ᵐ 08ˢ	55°	84%
Mullaittivu, Sri Lanka	+6	11.50	13.56	15.42	7ᵐ 30ˢ	55°	84%
Teknaf, Chittagong, Bangladesh	+6	12.45	14.33	16.32	7ᵐ 59ˢ	34°	84%
Sittwe/Akyab, Myamar (Burma)	+6.30	13.16	15.03	16.32	8ᵐ 36ˢ	33°	84%
Kyaukpyu, Myamar (Burma)	+6.30	13.16	15.03	16.32	4ᵐ 56ˢ	33°	84%
Zawngling, Mizoran, India	+5.30	12.19	14.05	15.33	5ᵐ 13ˢ	32°	84%
Chauk, Myamar (Burma)	+6.30	13.21	15.06	16.33	6ᵐ 59ˢ	32°	84%
Myingyan, Myamar (Burma)	+6.30	13.23	15.07	16.30	7ᵐ 28ˢ	32°	84%
Monywa, Myamar (Burma)	+6.30	13.24	15.07	16.34	8ᵐ 34ˢ	32°	84%
Mandalay, Myamar (Burma)	+6.30	13.26	15.09	10.34	7ᵐ 13ˢ	30°	83%
Mogok, Myamar (Burma)	+6.30	13.28	15.10	16.35	8ᵐ 24ˢ	28°	83%
Bhamo, Myamar (Burma)	+6.30	13.31	15.11	16.35	8ᵐ 04ˢ	27°	83%
RuIli, Yunnan, China	+8	15.02	16.42	18.06	8ᵐ 23ˢ	27°	83%
Myitkyina, Myamar (Burma)	+6.30	13.32	15.12	16.36	4ᵐ 37ˢ	26°	83%
Baoshan, Yunnan, China	+8	15.06	16.44	18.06	6ᵐ 29ˢ	24°	83%
Xiaguan (Dali), Yunnan, China	+8	15.08	16.45	18.06	8ᵐ 13ˢ	23°	83%
Dukou, Sichuan, China	+8	15.11	16.46	18.06	8ᵐ 08ˢ	21°	83%
Dongchuanm , Yohan, China	+8	15.15	16.48	18.07	3ᵐ 57ˢ	19°	83%
Xichang, Sichuan, China	+8	15.13	16.48	18.07	7ᵐ 08ˢ	20°	83%
Zhaotong, Sichuan, China	+8	15.16	16.48	18.07	7ᵐ 37ˢ	19°	83%
Zigong, Sichuan, China	+8	15.18	16.50	18.07	7ᵐ 10ˢ	19°	83%
Luzhou, Sichuan, China	+8	15.19	16.50	18.08	7ᵐ 57ˢ	17°	83%
Chongqing, Sichuan, China	+8	15.21	16.51	18.07	7ᵐ 52ˢ	15°	83%
Nanchong, Sichuan, China	+8	15.21	16.51	18.07	4ᵐ 39ˢ	14°	83%
Zhongxian, Chongqing, China	+8	15.23	16.52	18.05	7ᵐ 47ˢ	15°	83%
Wanxian, Chongqing, China	+8	15.24	16.52	18.03	7ᵐ 43ˢ	13°	83%
Enshi, Hubei, China	+8	15.26	16.52	17.59	6ᵐ 47ˢ	14°	83%
Yichang, Hubei, China	+8	15.28	16.53	17.51	4ᵐ 54ˢ	11°	83%
Shiyan, Hubei, China	+8	15.28	16.53	17.49	7ᵐ 01ˢ	10°	83%
Zhengzhou, Henan, China	+8	15.31	16.55	17.33	4ᵐ 28ˢ	7°	82%
Shangqiu, Henan, China	+8	15.33	16.55	17.25	7ᵐ 30ˢ	5°	82%
Laiwu, Shandong, China	+8	15.35	16.55	17.13	6ᵐ 11ˢ	4°	82%
Zibo, Shandong, China	+8	15.35	16.55	17.10	4ᵐ 37ˢ	3°	82%
Weifang, Shandong, China	+8	15.36	16.55	17.06	6ᵐ 32ˢ	2°	82%
Quindao, Shandong, China	+8	15.21	16.51	17.08	7ᵐ 52ˢ	1°	82%

2010 July 11, total

Total: French Polynesia, Easter Island, Chile
Partial: South Pacific, southern South America

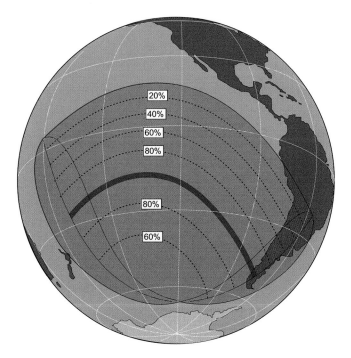

Course of the eclipse

1 End of eclipse at sunrise

2 Maximum at sunrise

3 Beginning of eclipse at sunrise

4 End of eclipse at sunset

5 Maximum at sunset

6 Beginning of eclipse at sunset

This eclipse begins in the south-western Pacific Ocean between New Zealand and Tonga, traverses the South Pacific in a northerly arch and ends at the southern tip of South America in Patagonia. The maximum totality length of 5 minutes 20 seconds puts it among the longest eclipses. On its path across the vast, empty Pacific Ocean, Easter Island is centrally under its shadow.

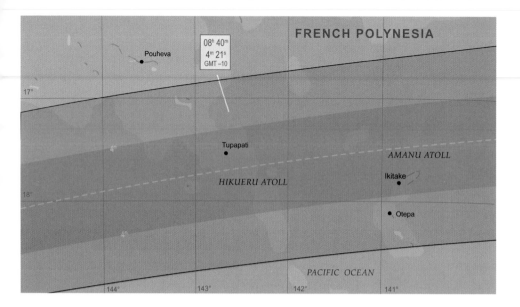

FRENCH POLYNESIA

08ʰ 40ᵐ
4ᵐ 21ˢ
GMT −10

Pouheva

17°

Tupapati

AMANU ATOLL

HIKUERU ATOLL

Ikitake

18°

Otepa

PACIFIC OCEAN

144°   143°   142°   141°

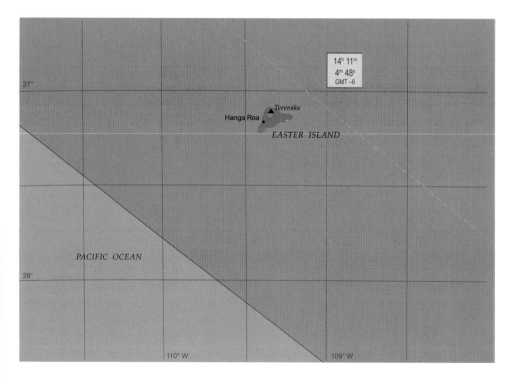

14ʰ 11ᵐ
4ᵐ 48ˢ
GMT −6

27°

Terevaka

Hanga Roa

EASTER ISLAND

PACIFIC OCEAN

28°

110° W   109° W

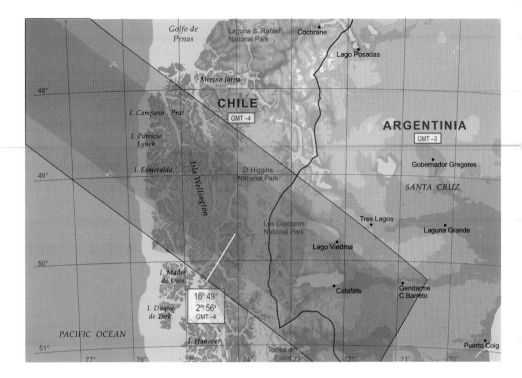

This island, named after the day of its discovery at Easter 1722, has become famous because of its monumental sculptures made of volcanic stone, the moai. This isolated island, formed from three extinct volcanoes which give the island its distinctive form, was populated in the fifth century by Polynesians. Anyone considering a trip to this gem in the Pacific should go in July 2010. A solar eclipse in front of the silhouettes of the 'silent giants' must surely rank with the Egyptian eclipse of 1908 in terms of its primal and elemental expressive power. Alternatively, observing the eclipse from the island's 500 m high volcanic cone Terevaka, surrounded by the Pacific, will be a memorable experience.

At Easter Island, called by its inhabitants 'Rapa Nui' and 'Te Pito O Te Henna' (navel of the world), totality will last for 4 minutes and 44 seconds with the 186 km wide shadow covering the whole of the island.

The eclipse ends after crossing the 3500 m high southern Andes before reaching western Santa Cruz in Argentina at

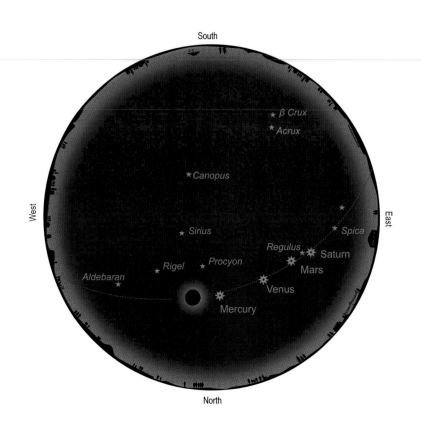

sunset. In the second minute of totality, the Sun is only 2° above the horizon.

The stars during the eclipse

From Patagonia, the darkened Sun is seen shortly before sunset. Hence an unobstructed view towards the west is required. Mercury is visible to the right of the Sun. Venus, Mars and Saturn follow at an angle upwards from the horizon. Whereas Venus is immediately visible because of its brightness, Mars and Saturn cannot be seen so easily.

Weather

Easter Island has a subtropical climate, which means that July and August are the coolest months of the year. Temperatures fall to about 8°C at night. Although with 1130 mm of annual precipitation and July falling in the wet season, loose cloud cover can nevertheless be expected. Even with a cloudy sky there can be few more impressive panoramic landscapes during a solar eclipse than the silent moais of the mysterious Easter Island culture.

The central eclipse ends in southern Patagonia. This harsh, open landscape may bring a particularly impressive experience of nature. Here the shadow corridor crosses the Los Glaciares National Park. Both the protected Magellan rain forest in the west and the numerous glaciers in the east turn this landscape 'at the ends of the Earth' into a magnificent experience of nature. The stunning language of a solar eclipse will chime together particularly well with the powerful landscape.

2010 July 11, Sunday Location	Time zone (GMT ±)	Begin-ning	Maxi-mum	End	Dura-tion	Height of Sun
150 km S of Tepati, Tahiti, Fr. Pol.	−10	07.18	08.30	09.53	3ᵐ 52ˢ	24°
Tupapati, Hikueru, Fr. Polynesia	−10	07.22	08.39	10.07	4ᵐ 17ˢ	31°
Ikitake, Amuna Atoll, Fr. Pol.	−10	07.24	08.43	10.13	4ᵐ 12ˢ	33°
Hanga Roa, Easter Island, Chile	−6	12.41	14.11	15.34	4ᵐ 44ˢ	39°
O'Higgins Nat.Park (South), Chile	−4	15.44	16.50	17.06	2ᵐ 28ˢ	2°
Lago Viedma, S. Cruz, Argentina	−3	16.45	17.50	17.59	2ᵐ 24ˢ	1°
Calafate, Santa Cruz, Argentina	−3	16.44	17.50	17.58	2ᵐ 47ˢ	1°

2012 May 20/21, annular

Annular: Southern China, southern Japan, North Pacific, USA (California to New Mexico)
Partial: Eastern Asia, North Pacific, North America (except east)

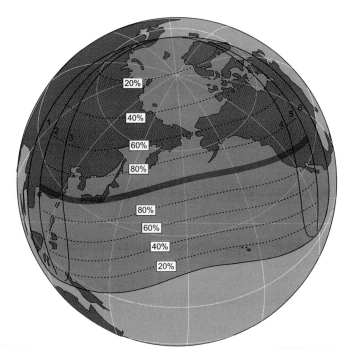

Course of the eclipse

1 End of eclipse at sunrise

2 Maximum at sunrise

3 Beginning of eclipse at sunrise

4 End of eclipse at sunset

5 Maximum at sunset

6 Beginning of eclipse at sunset

While 2011 saw only partial eclipses, we now have a central eclipse stretching from southern China to the southern states of USA. The shadow first touches the ground in the morning of May 21 on the south coast of China where the cities of Guangzhou (Canton) and Hong Kong are crossed. Then the path of the

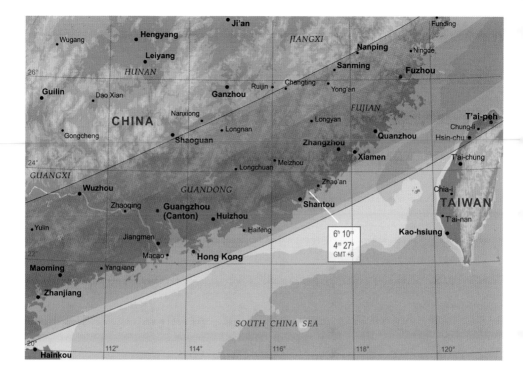

eclipse travels across the East China Sea and touches the northern tip of Taiwan so that in the capital T'ai-peh 2 minutes of annular shadow can be experienced. Then the whole of the south-eastern part of Japan is affected although the metropoles of Osaka, Kyoto and Nagoya are positioned near the edge of the band so that the length of the eclipse is only 2 to 3 minutes. In contrast, Tokyo lies right at the centre and thus enjoys the annular eclipse for over 5 minutes.

The eclipse travels on over the Pacific and passes so far north that it almost touches the Andreanof Islands, part of the Alaskan Aleutian Islands. Having crossed the international date line, it reaches the mainland on the evening of May 20 in northern California, moving on to cross the states of Nevada, Utah, Arizona, Colorado and New Mexico. Philip Harrington, who has written an informative book about the observation of eclipses, points out that alongside the Grand Canyon several magnificent natural monuments and nature parks are in this eclipse zone. Together

2012 May 21, Monday Location	Time zone (GMT ±)	Begin- ning	Maxi- mum	End	Dura- tion	Height of Sun	Cover- age
Zhanjiang, Guangdong, China	+8	06.02	06.09	07.14	3m 55s	1°	87%
Maoming, Guangdong, China	+8	05.59	06.09	07.15	4m 10s	4°	87%
Yangjiang, Guangdong, China	+8	05.55	06.09	07.15	4m 03s	3°	87%
Wuzhou, Guangxi, China	+8	05.55	06.09	07.17	2m 22s	3°	87%
Zhaoqing, Guangdong, China	+8	05.51	06.10	07.17	4m 10s	4°	87%
Macau, China	+8	05.48	06.10	07.16	3m 42s	6°	87%
Guangzhou (Canton), China	+8	05.47	06.10	07.17	4m 22s	6°	87%
Hong Kong, China	+8	05.45	06.10	07.16	3m 27s	5°	87%
Longchuan, Guangdong, China	+8	05.38	06.10	07.19	4m 21s	6°	87%
Meizhou, Guangdong, China	+8	05.34	06.11	07.20	4m 27s	8°	87%
Longyan, Fujian, China	+8	05.27	06.11	07.23	2m 22s	10°	87%
Zhangzhou, Fujian, China	+8	05.27	06.11	07.21	4m 26s	5°	87%
Xiamen, Fujian, China	+8	05.26	06.11	07.21	4m 15s	10°	87%
Yong'an, Fujian, China	+8	05.26	06.11	07.23	2m 26s	9°	87%
Quanzhou, Fujian, China	+8	05.23	06.11	07.22	4m 27s	10°	87%
Fuzhou, Fujian, China	+8	05.18	06.12	07.24	4m 14s	12°	87%
Ningole, Fujian, China	+8	05.16	06.12	07.25	3m 23s	12°	87%
Hsin-chu, Taiwan	+8	05.14	06.12	07.23	1m 40s	11°	87%
Chung-li, Taiwan	+8	05.13	06.13	07.23	1m 56s	13°	87%
T'ai-peh, Taiwan	+8	05.11	06.13	07.24	1m 43s	13°	87%
Toshima, Nansei, Japan	+9	06.11	07.20	08.39	4m 14s	23°	88%
Yaku, Nansei, Japan	+9	06.11	07.21	08.40	4m 20s	24°	88%
Kukinaga, Tanega-shima, Japan	+9	06.11	07.21	08.41	4m 17s	24°	88%
Kagoshima, Kyushu, Japan	+9	06.13	07.22	08.43	4m 16s	25°	88%
Miyazaki, Kyushu, Japan	+9	06.13	07.23	08.44	4m 29s	25°	88%
Uwajima, Shikoku, Japan	+9	06.15	07.25	08.48	2m 38s	27°	88%
Kochi, Shikoku, Japan	+9	06.15	07.26	08.50	3m 16s	28°	88%
Wakayama, Japan	+9	06.16	07.28	08.53	3m 50s	30°	88%
Kobe, Japan	+9	06.17	07.29	08.54	1m 40s	30°	88%
Osaka, Japan	+9	06.17	07.29	08.54	2m 51s	30°	88%
Kyoto, Japan	+9	06.18	07.29	08.55	1m 30s	31°	88%
Nagoya, Japan	+9	06.18	07.31	08.57	3m 43s	32°	88%
Gifu, Japan	+9	06.18	07.31	08.58	2m 17s	32°	88%
Hamamatsu, Japan	+9	06.17	07.31	08.58	4m 59s	33°	88%
Shizuoka, Japan	+9	06.18	07.32	09.00	5m 02s	33°	88%
Numazu, Japan	+9	06.18	07.33	09.00	5m 02s	33°	88%

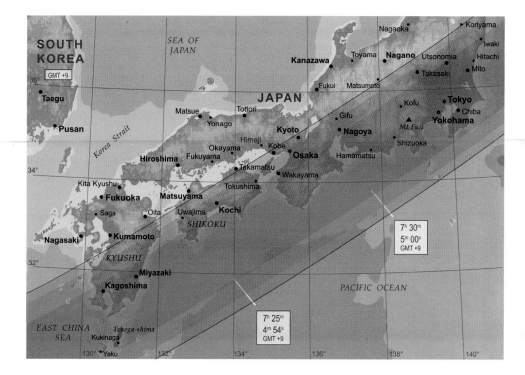

with the low annular eclipsed Sun, they produce a wonderful correspondence between the sublime celestial spectacle and the grandiose landscape. This includes, for example, the Lassen Volcanic National Park which contains four inactive volcanoes over 3000 m (10 000 ft) high. Hot springs and lava shapes are a reminder of this geological activity. Lake Tahoe, set beautifully in

2012 May 21, Monday Location	Time zone (GMT ±)	Begin- ning	Maxi- mum	End	Dura- tion	Height of Sun	Cover- age
Mt Fuji, Japan	+9	06.18	07.33	09.01	4m 58s	33°	88%
Kofu, Japan	+9	06.19	07.33	09.01	4m 34s	33°	88%
Yokohama, Japan	+9	06.19	07.35	09.03	5m 04s	35°	88%
Tokyo, Japan	+9	06.19	07.35	09.03	5m 04s	35°	88%
Takasaki, Japan	+9	06.20	07.35	09.03	3m 33s	35°	88%
Chiba, Japan	+9	06.19	07.35	09.03	5m 01s	35°	88%
Mito, Japan	+9	06.20	07.36	09.05	4m 56s	36°	88%
Iwaki, Japan	+9	06.22	07.38	09.07	4m 16s	37°	88%

2012 May 20, Sunday Location	Time zone (GMT ±)	Begin-ning	Maxi-mum	End	Dura-tion	Height of Sun	Cover-age
Gold Beach, Oregon, USA	−7	17.09	18.26	19.35	4m 04s	22°	88%
Brookings, Oregon, USA	−7	17.08	18.27	19.35	4m 32s	22°	88%
Crescent City, California, USA	−7	17.08	18.27	19.36	4m 44s	22°	88%
Eureka, California, USA	−7	17.10	18.27	19.37	3m 59s	22°	88%
Grants Pass, Oregon, USA	−7	17.08	18.27	19.34	2m 42s	22°	88%
Medford, Oregon, USA	−7	17.09	18.27	19.35	2m 31ss	21°	88%
Weaverville, California, USA	−7	17.11	18.28	19.37	4m 35s	21°	88%
Dunsmuir, California, USA	−7	17.11	18.28	19.36	4m 37s	21°	88%
Redding, California, USA	−7	17.12	18.29	19.37	4m 36s	20°	88%
Lassen Peak, California, USA	−7	17.13	18.29	19.37	4m 42s	20°	88%
Chico, California, USA	−7	17.14	18.29	19.38	3m 24s	19°	88%
Alturas, California, USA	−7	17.12	18.29	19.35	2m 30s	19°	88%
Emigrant Gap, California, USA	−7	17.15	18.30	19.38	3m 22s	19°	88%
Susanville, California, USA	−7	17.14	18.30	19.37	4m 40s	19°	88%
Doyle, California, USA	−7	17.15	18.30	19.37	4m 41s	18°	88%
Reno, Nevada, USA	−7	17.16	18.30	19.38	4m 27s	18°	88%
South Lake Tahoe, California	−7	17.16	18.31	19.38	3m 04s	17°	88%
Carson City, Nevada, USA	−7	17.16	18.31	19.38	3m 52s	17°	88%
Fernley, Nevada, USA	−7	17.16	18.31	19.37	4m 38s	17°	88%
Lovelock, Nevada, USA	−7	17.16	18.31	19.36	4m 08s	17°	88%
Austin, Nevada, USA	−7	17.18	18.32	19.37	4m 20s	15°	88%
Tonopah, Nevada, USA	−7	17.20	18.32	19.38	3m 29s	15°	88%
Warm Springs, Nevada, USA	−7	17.20	18.33	19.38	4m 19s	14°	88%
Currant, Nevada, USA	−7	17.20	18.33	19.37	4m 27s	14°	88%
Ely, Nevada, USA	−7	17.20	18.33	19.36	3m 09s	14°	88%
Baker, Nevada, USA	−7	17.20	18.33	19.36	2m 59s	13°	88%
Caliente, Nevada, USA	−7	17.22	18.34	19.38	4m 24s	12°	88%
Milford, Utah, USA	−6	18.22	19.34	20.36	3m 32s	11°	88%
St George, Utah, USA	−6	18.23	19.34	20.36	4m 14s	11°	88%
Cedar City, Utah, USA	−6	18.23	19.34	20.35	4m 29s	12°	88%
Kanab, Utah, USA	−6	18.24	19.34	20.31	4m 29s	11°	88%
Tropic, Utah, USA	−6	18.23	19.34	20.31	4m 13s	11°	88%
Grand Canyon Village, Arizona	−7	17.26	18.35	19.27	3m 16s	10°	88%

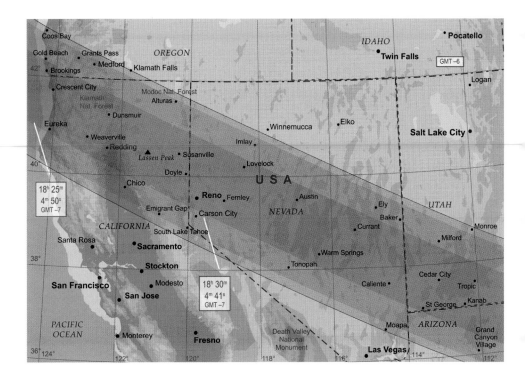

the Sierra Nevada, and the sparsely populated desert of Nevada lie in the shadow's path before it comes to the colourful desert landscape of the Grand Canyon in Arizona. Seeing the eclipse shadow approaching from the west in the late afternoon over this bizarre landscape will leave a deep impression. Passing over Albuquerque it ends at sunset in Lubock, Texas.

The weather

In Japan, and even more so on the Chinese coast, thicker cloud cover is to be expected in May. The dry landscape of northern California and the deserts of Nevada, Utah and Arizona are known for their sunny days. A clear sky, or at worst, only slight cloud cover, can be expected on the west coast of California. Further inland the likelihood of cloud is greatly reduced.

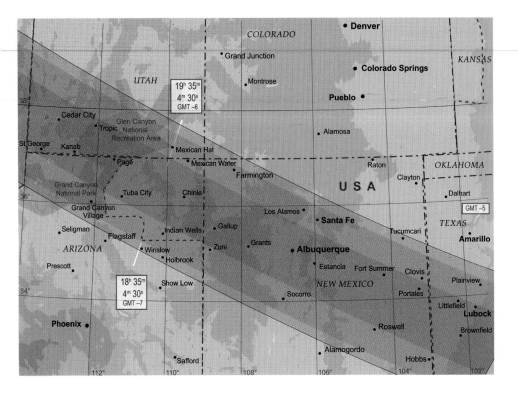

SOLAR ECLIPSES

2012 May 20, Sunday Location	Time zone (GMT ±)	Begin-ning	Maxi-mum	End	Dura-tion	Height of Sun	Cover-age
Tuba City, Arizona, USA	−6	18.26	19.35	20.24	4m 06s	9°	87%
Mexican Hat, Utah, USA	−6	18.25	19.35	20.21	3m 35s	9°	87%
Mexican Water, Arizona , USA	−6	18.25	19.35	20.20	3m 49s	9°	87%
Indian Wells, Arizona , USA	−6	18.27	19.35	20.18	3m 37s	8°	87%
Chinle, Arizona , USA	−6	18.26	19.35	20.17	4m 29s	8°	87%
Farmington, New Mexico, USA	−6	18.26	19.36	20.13	3m 02s	7°	87%
Gallup, New Mexico, USA	−6	18.28	19.36	20.13	4m 23s	7°	87%
Zuni, New Mexico, USA	−6	18.28	19.36	20.12	3m 52s	7°	87%
Grants, New Mexico, USA	−6	18.28	19.36	20.08	4m 22s	6°	87%
Los Alamos, New Mexico, USA	−6	18.28	19.36	20.04	3m 19s	6°	87%
Santa Fe, New Mexico, USA	−6	18.28	19.36	20.02	3m 30s	7°	87%
Albuquerque, New Mexico, USA	−6	18.29	19.36	20.03	4m 24s	5°	87%
Estancia, New Mexico, USA	−6	18.29	19.36	20.02	4m 23s	5°	87%
Fort Summer, New Mexico, USA	−6	18.30	19.37	19.52	4m 12s	3°	87%
Roswell, New Mexico, USA	−6	18.31	19.37	19.51	3m 45s	3°	87%
Clovis, New Mexico, USA	−6	18.30	19.37	19.48	3m 51s	3°	87%
Portales, New Mexico, USA	−6	18.30	19.37	19.48	4m 08s	3°	87%
Plainsview, Texas, USA	−5	19.31	20.37	20.41	3m 07s	1°	87%
Lubbock, Texas, USA	−5	19.31	20.37	20.41	4m 09s	1°	87%

2012 November 13/14, total

Total: Northern Australia, South Pacific
Partial: Southern Pacific, Antarctica, southern South America

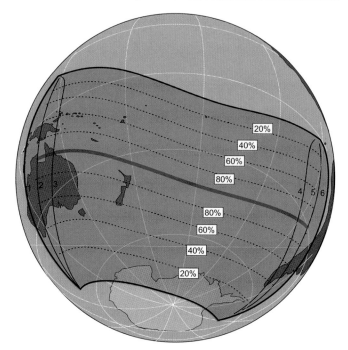

Course of the eclipse

1 End of eclipse at sunrise

2 Maximum at sunrise

3 Beginning of eclipse at sunrise

4 End of eclipse at sunset

5 Maximum at sunset

6 Beginning of eclipse at sunset

The solar eclipse of November 13/14, will mainly move across the South Pacific. The eclipse starts at sunrise on November 14 in Arnhem Land, Northern Territory, Australia, east of the Kakadu National Park. This region, named after an Aboriginal tribe, is a world heritage site. Extensive forests as well as sandstone highlands give the landscape its character. The shadow corridor, which at the start is about 120 km wide, then travels eastwards

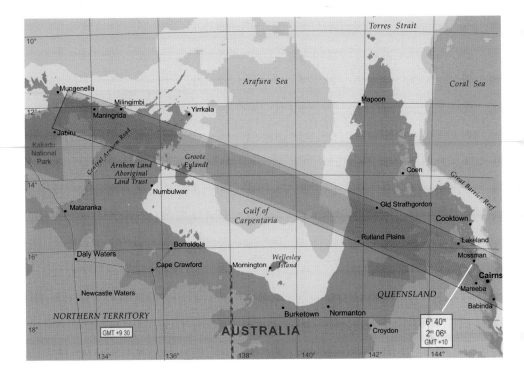

across Arnhem Land. Forests and steep rock formations mark this region with its tropical character. The Aborigines here can look back on thousands of years of history documented in beautiful rock paintings. Arnhem Land is Aborginal Land and is closed to independent travellers. However, the Central Arnhem Road is open by permit.

The totality zone passes the island of Groote Eylandt, crosses the Gulf of Carpentaria, and reaches northern Queensland where the city of Cairns is darkened. Cairns is the gate to the Great Barrier Reef — the world's greatest coral belts with a length of 2000 km.

South of New Caledonia the eclipse traverses the Tropic of Capricorn. At maximum, the total eclipse has a duration of 4 minutes 2 seconds.

The eclipse ends at sunset on November 13 about 1000 km west of the Chilean coast and 400 km north of the Juan Fernandez Islands which have become famous as the islands to inspire Defoe's book, *Robinson Crusoe*.

Weather

In north-eastern Australia there is almost constant open, loose to medium cloud cover. With luck, observation of the obscured Sun should therefore be possible. The best conditions are likely to be on the east coast.

2012 Nov 14, Wednesday Location	Time zone (GMT ±)	Begin- ning	Maxi- mum	End	Dura- tion	Height of Sun
Maningrida, N.T., Australia	+9.30	06.02	06.06	07.01	1m 26s	0.5°
Central Arnhem Road Lat. 13° S	+9.30	05.57	06.07	07.02	1m 30s	2°
Rutland Plains, Qld., Australia	+10	05.56	06.38	07.38	1m 31s	10°
Lakeland, Queensland, Australia	+10	05.44	06.39	07.39	1m 06s	13°
Mossman, Queensland, Australia	+10	05.44	06.40	07.40	1m 58s	13°
Mareeba, Queensland, Australia	+10	05.45	06.40	07.40	1m 53s	14°
Cairns, Queensland, Australia	+10	05.45	06.39	07.41	2m 04s	14°
Babinda, Queensland, Australia	+10	05.45	08.40	07.41	1m 38s	14°

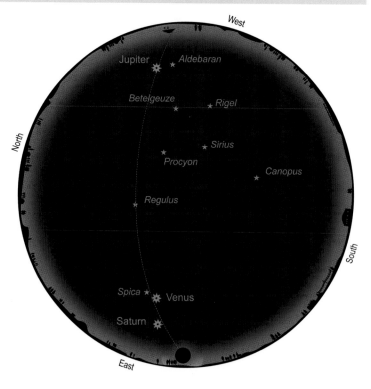

Venus and perhaps Saturn are visible above the Sun while Jupiter is setting in the west.

2013 May 10, annular

Annular: Australia, Papua New Guinea, Solomon Islands, Kiribati
Partial: Indonesia, Australia, northern New Zealand, Polynesia, souther Pacific Ocean

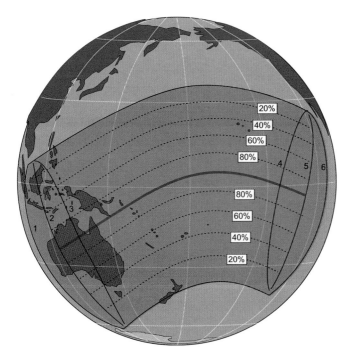

Only six months after the total solar eclipse of November 13, 2012, another central eclipse passes across Australia. On this occasion the New Moon and its apogee lie only three days apart, so it is not big enough to cover the Sun completely, resulting in an annular eclipse in which 8.5% of the Sun remains uncovered.

Course of the eclipse

1 End of eclipse at sunrise

2 Maximum at sunrise

3 Beginning of eclipse at sunrise

4 End of eclipse at sunset

5 Maximum at sunset

6 Beginning of eclipse at sunset

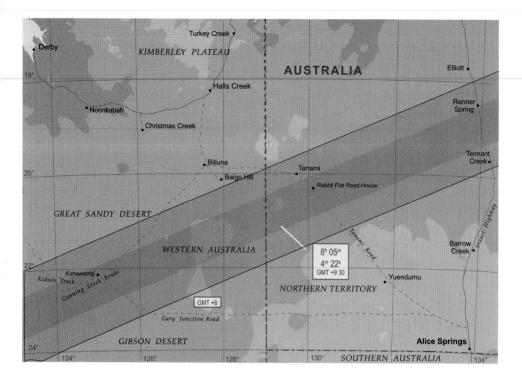

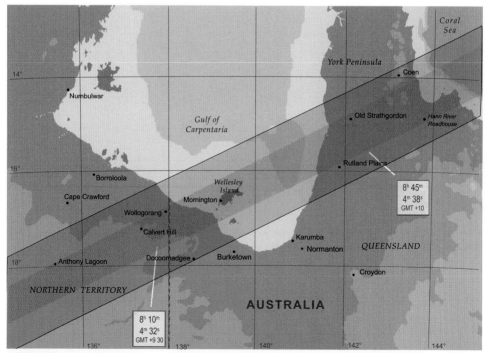

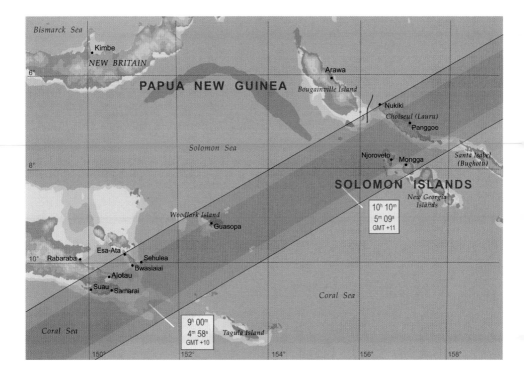

The band of the eclipse starts in the Great Sandy Desert in Western Australia. The shadow band travels north-eastwards through vast uninhabited parts of the outback, crossing the Stuart Highway between Tennant Creek and Renner Spring. In the Gulf of Carpentaria, Wellesley Island lies at the centre of the corridor. Then the eclipse traverses the York Peninsula crossing the path of the previous year's eclipse. Here the speed of the shadow is 5000 km/h and the eclipse again crosses the Great Barrier Reef before reaching the Coral Sea.

After an 800 km sweep across the Coral Sea, the shadow band touches the south-eastern tip of New Guinea as well as Woodlark Island and the Solomon Islands. Just south of the Equator, the small island of Nauru is traversed. Kiribati's Gilbert Islands are also crossed by the shadow. The eclipse ends at sunset on May 9 about 5000 km off the Peruvian coast.

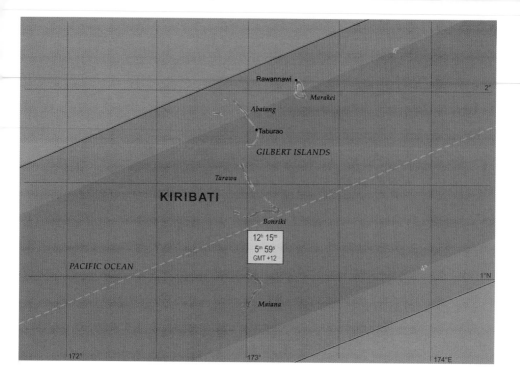

Weather

In the Australian desert, there is about 70% probability of clear or only slightly cloudy skies. On the east coast, the conditions deteriorate somewhat. New Guinea has a hot and humid tropical climate. The rainy season ends in March.

The Solomon Islands, situated further east, have a semi-tropical climate and the rainy season ends in April.

2013 May 10, Friday Location	Time zone (GMT ±)	Begin- ning	Maxi- mum	End	Dura- tion	Height of Sun	Cover- age
Kunawarritji, West Australia	+8	06.07	06.34	07.51	4ᵐ 11ˢ	6°	89%
Rabbit Flat Roadho. NT, Aus	+9 30	07.13	08.05	09.27	4ᵐ 02ˢ	11°	89%
Tennant Creek, NT, Australia	+9 30	06.56	08.07	09.33	2ᵐ 47ˢ	16°	89%
Renner Spring, NT, Australia	+9 30	06.55	08.08	09.33	3ᵐ 43ˢ	16°	89%
Anthony Lagoon, NT, Australia	+9 30	06.55	08.09	09.36	4ᵐ 15ˢ	18°	89%
Calvert Hills, NT, Australia	+9 30	06.56	08.10	09.39	4ᵐ 24ˢ	20°	90%
Wollogorang, NT, Australia	+9 30	06.56	08.11	09.41	4ᵐ 32ˢ	20°	90%
Docoomadgee, Qld. Australia	+10	07.26	08.41	10.11	1ᵐ 29ˢ	21°	90%
Rutland Plains, Qld. Australia	+10	07.27	08.44	10.19	4ᵐ 35ˢ	25°	90%
Old Strathgordon, Qld. Australia	+10	07.27	08.46	10.21	3ᵐ 47ˢ	26°	90%
Coen, Queensland. Australia	+10	07.28	08.48	10.24	2ᵐ 49ˢ	28°	90%
Hann River Roadhouse, Qld.	+10	07.28	08.47	10.25	2ᵐ 30ˢ	28°	90%
Suau, Papua New Guinea	+10	07.32	08.59	10.45	4ᵐ 05ˢ	37°	90%
Samarai, Papua New Guinea	+10	07.33	09.00	10.47	4ᵐ 54ˢ	39°	90%
Guasopo, Papua New Guinea	+10	07.35	09.05	10.56	4ᵐ 49ˢ	42°	90%
Njoroveto, Solomon Islands	+11	08.39	09.14	12.16	4ᵐ 29ˢ	48°	91%
Panggoe, Solomon Islands	+11	08.40	10.16	12.14	5ᵐ 12ˢ	50°	91%
Taburao, Gilbert Is. Kiribati	+12	10.16	12.16	14.23	4ᵐ 59ˢ	74°	91%

2013 November 3, hybrid (annular-total)

Annular: Atlantic, Gabon, Congos, Uganda, Kenya
Partial: northern South America, Newfoundland, northern and central Atlantic, Spain, Africa, Arabian Peninsula

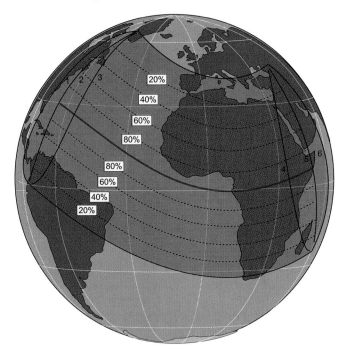

Course of the eclipse

1 End of eclipse at sunrise

2 Maximum at sunrise

3 Beginning of eclipse at sunrise

4 End of eclipse at sunset

5 Maximum at sunset

6 Beginning of eclipse at sunset

This eclipse starts off the East Coast of America, south-west of Bermuda, so no more than the conclusion of the eclipse can be observed from Florida, the Carolinas and Virginia at sunrise. Only

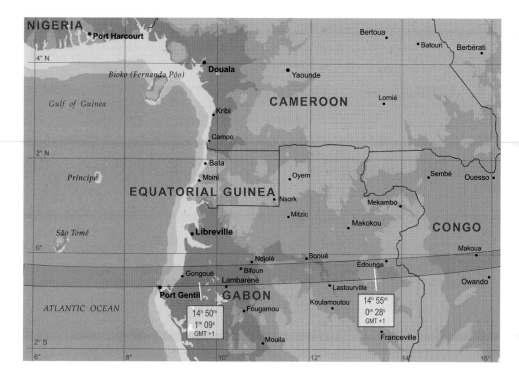

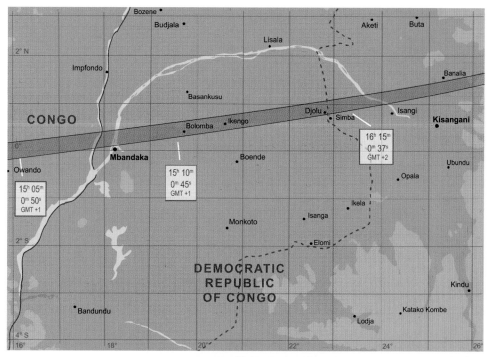

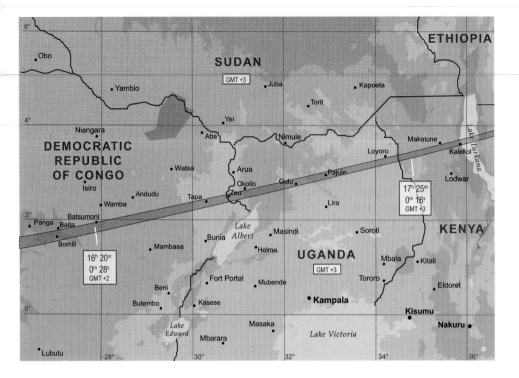

30% of the Sun is covered at that point. At first the eclipse is annular but after having travelled less than 500 km the moon's umbra reaches the Earth's surface. The path of the eclipse, very small at the beginning, grows as it travels across the Atlantic to become almost 60 km wide giving a maximum duration of the eclipse of 1 minutes 40 seconds. The eclipse passes the Cape Verde Islands 450 km to the south so that from these islands the coverage is 84% of the Sun. The core shadow then reaches the African continent with a width of only 45 km at the small town of Gongué in Gabon, a country largely covered by tropical rain forest. The eclipse travels across the hilly interior and then into the highlands of Congo. After crossing the Congo (Zaire) River it touches the town of Mbandaka in the Democratic Republic of Congo.

In Uganda the eclipse only lasts 22 seconds and its track has narrowed to 18 km. The eclipse peters out in northern Kenya at Lake Turkana where it appears again as an annular eclipse at its finish.

Weather

In Gabon, the Congos and Uganda, one of the two rainy seasons lasts from October to November. In this period there is frequent and sustained rain. Although Uganda is also situated around the equator, the temperatures are milder and the chances of finding a patch of clear sky are better. However, the eclipse is only visible here for a relatively short period of time (20 seconds or less). All in all the conditions for this African eclipse are rather unfavourable. One attractive feature is the proximity of Saturn and Mercury. Both are positioned a bare 4° alongside the obscured Sun.

2013 Nov 3, Sunday Location	Time zone (GMT ±)	Begin-ning	Maxi-mum	End	Dura-tion	Height of Sun
Gongoué, Gabon	+1	13.13	14.51	16.15	0ᵐ 55ˢ	47°
Bifoun, Gabon	+1	13.17	14.54	16.16	0ᵐ 46ˢ	45°
20 km S of Booué, Gabon	+1	13.21	14.57	16.18	1ᵐ 01ˢ	43°
Edounga, Gabon	+1	13.27	15.01	16.20	0ᵐ 25ˢ	40°
Makoua, Congo	+1	13.32	15.04	16.21	0ᵐ 47ˢ	37°
Bolomba, Dem. Rep. Congo	+1	13.42	15.10	16.24	0ᵐ 44ˢ	32°
Djolu, Dem. Rep. Congo	+1	13.49	15.14	16.26	0ᵐ 14ˢ	29°
Bomili, Dem. Rep. Congo	+2	14.59	16.19	17.27	0ᵐ 30ˢ	23°
Zeu, Uganda	+3	16.06	17.22	18.28	0ᵐ 21ˢ	18°
Gulu, Uganda	+3	16.08	17.23	18.28	0ᵐ 16ˢ	16°
Pajule, Uganda	+3	16.09	17.24	18.27	0ᵐ 18ˢ	15°
Kalekol, Kenya	+3	16.13	17.26	18.16	0ᵐ 15ˢ	12°

2014 April 29, annular

Annular: Antarctica
Partial: Indian Ocean, Australia.

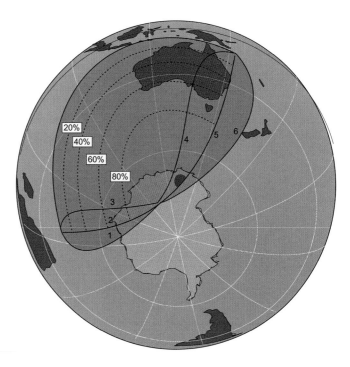

Course of the eclipse

1 End of eclipse at sunrise

2 Maximum at sunrise

3 Beginning of eclipse at sunrise

4 End of eclipse at sunset

5 Maximum at sunset

6 Beginning of eclipse at sunset

This eclipse in the first half of 2014 is the first within Saros cycle 148 to become central. All previous members of this eclipse series produced partial eclipses visible in Antarctica. Now the umbra reaches the Earth in the course of its northward progress. But since the core shadow only grazes the Earth, this eclipse does not follow a path as such but gives rise to a point-like shadow zone in Wilkes Land in the Antarctic. At the time of the eclipse the Sun has only risen 2° above the horizon.

2015 March 20, total

Total: North Atlantic, Faeroes, Spitsbergen, Arctic Ocean.
Partial: Europe, North Africa, North-west Asia.

Course of the eclipse

1 End of eclipse at sunrise

2 Maximum at sunrise

3 Beginning of eclipse at sunrise

4 End of eclipse at sunset

5 Maximum at sunset

6 Beginning of eclipse at sunset

At the spring equinox in the northern hemisphere, a total solar eclipse occurs whose core shadow line is very wide, averaging 430 km. It starts east of Newfoundland and then circles around Greenland and Iceland on its way through the Norwegian Sea. The totality runs through Rockall; west of St Kilda cover is

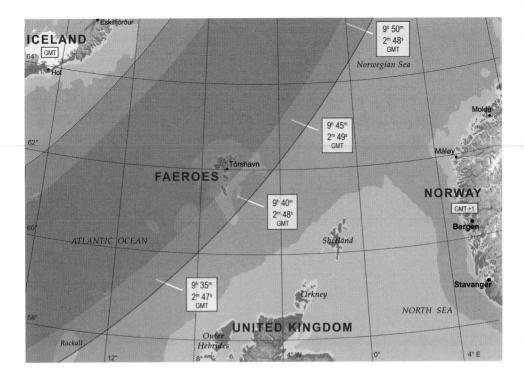

98.4%. The Faeroe Islands lie in the core area and thus enjoy a totality of about 2 minutes 10 seconds. From its peaks, such as Slaettaratindur (882 m) on Eysturoy, the rugged group of islands with its relentless north Atlantic winds will provide a grandiose view of the eclipse, approaching at a speed of 3200 km/h from the south-west.

The core shadow line moves on northwards, 250 km off the Norwegian coast. It then reaches Spitsbergen, the Norwegian archipelago in the Arctic Ocean, covering the whole of Svalbard (Prins Karls Forland, Bear Island, Edgeøya, Barentsøya, Nordausth-East Land) with the exception of the most easterly islands. The settlements of this island group, which has only been inhabited since the late nineteenth century, are almost all situated on the north side of the southern island. There are no tourist hotels on Spitsbergen. Visitors must find accommodation in private houses or stay in a spartan campsite. Visitors arriving inadequately equipped will be required to leave.

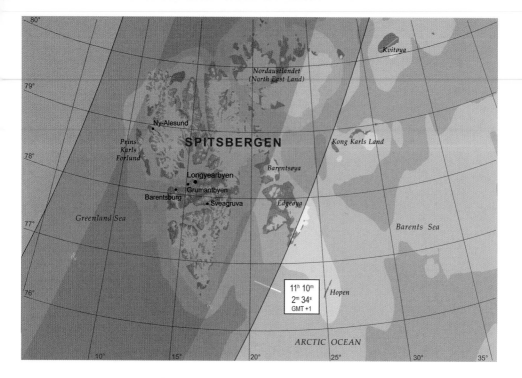

An unobstructed view towards the south-west is required to observe the eclipse. Together with the impressive ice landscape of these islands, a solar eclipse seen from this northern viewpoint is something special: since the Sun is angled no more than 11° above the horizon in this location halfway between the North Cape and the North Pole, the Moon moves horizontally into the Sun.

The further north we go, the lower the obscured Sun stands in the sky. At the North Pole itself, it will be directly on the horizon because the eclipse occurs at the spring equinox, the day on which the Sun can be seen above the horizon once again after six months of darkness.

Weather

The Faeroe Islands are known for their constant rain and thick cloud cover. Thus in March there is only 15 to 20% probability of a clear sky or light cloud. It is very likely (80%) that there will be

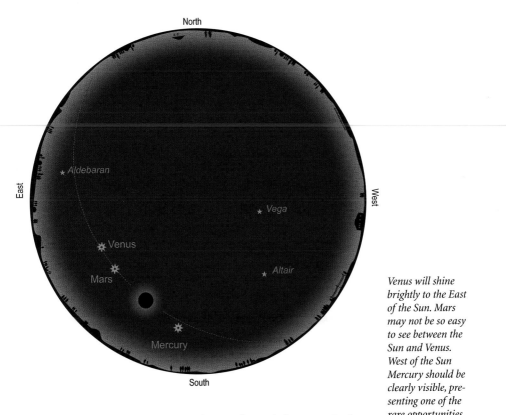

North

East

West

★ Aldebaran

★ Vega

✹ Venus

✹ Mars

★ Altair

✹ Mercury

South

Venus will shine brightly to the East of the Sun. Mars may not be so easy to see between the Sun and Venus. West of the Sun Mercury should be clearly visible, presenting one of the rare opportunities to see this elusive planet so far in the north.

thick cloud cover. Prospects are better for Spitsbergen. A clear view or only a few clouds in the sky can be expected with a probability of 30 to 35%.

2015 March 20, Friday Location	Time zone (GMT ±)	Begin-ning	Maxi-mum	End	Dura-tion	Height of Sun
Rockall (Atlantic)	0	8.28	9.31	10.36	1ᵐ 06ˢ	19°
Tórshavn, Faeroes	0	8.39	9.42	10.47	2ᵐ 17ˢ	20°
Longyearbyen, Spitsbergen	+1	10.12	11.12	12.12	2ᵐ 31ˢ	11°

2016 March 9, total

Total: Indonesia, Pacific Ocean
Partial: Northern and western Australia, Eastern Asia, northern Pacific to western Alaska

Course of the eclipse

1 End of eclipse at sunrise

2 Maximum at sunrise

3 Beginning of eclipse at sunrise

4 End of eclipse at sunset

5 Maximum at sunset

6 Beginning of eclipse at sunset

The eclipse begins in the Indian Ocean and first reaches land at the islands of Pagai Utara and Pagai Selatan off Indonesia. At totality the Sun from this vantage point has already reached 12° and remains covered for almost 2 minutes. The shadow then travels across southern Sumatra, covering Palembang, crossing

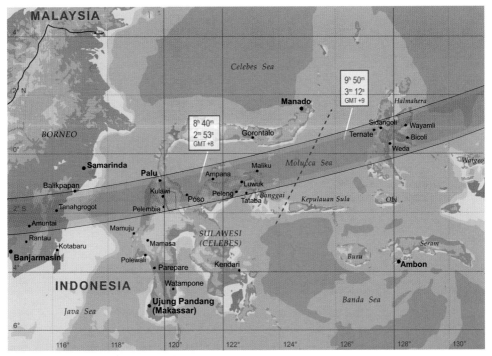

the Java Sea and traversing the south part of Bangka, the islands of Belitung, Borneo and Sulawesi (Celebes) in Indonesia. Here the eclipse has grown to a length of 3 minutes. All the islands have a tropical climate with high rainfall throughout the year. Halmahera is the last island traversed by the totality before the eclipse leaves the Indonesian archipelago and crosses the Pacific.

The shadow band passes about 1000 km to the north of Hawaii before ending 200 km off the North American west coast.

Seen from Hawaii, the eclipse has 62% cover. The Sun is positioned 12° above the horizon so that an unrestricted view towards the west will show an impressive upright sickle of the Sun above the ocean.

Weather

Sumatra and Borneo have a tropical climate and are characterized by frequent and sustained rainfall. In March, the probability of a clear or partially clear sky is as low as 10%. The chances are slightly more favourable on Sulawesi (Celebes) Island. If the opportunities for observing the eclipse are unfavourable in a tropical location, compensation is provided by the wealth of nature in these latitudes: the tropical plant and animal worlds are by their nature particularly closely aligned with the Sun. The much described phenomenon that the vitality of nature is curiously interrupted during totality will no doubt occur in a particularly intense and dramatic way in such an environment.

2016 March 9, Wednesday Location	Time zone (GMT ±)	Beginning	Maximum	End	Duration	Height of Sun
Buriai, Pegai Selatan, Indonesia	+7	06.28	07.20	08.26	1^m 40^s	13°
Mukomuko, Sumatra, Indonesia	+7	06.25	07.20	08.27	1^m 37^s	14°
Lubuk Linggau, Sumatra, Indon'	+7	06.21	07.21	08.29	1^m 17^s	16°
Kasmaran, Sumatra, Indonesia	+7	06.21	07.21	08.30	1^m 58^s	16°
Palembang, Sumatra, Indonesia	+7	06.21	07.22	08.32	2^m 00^s	18°
Bedinggong, Bangka, Indonesia	+7	06.21	07.23	08.34	2^m 07^s	19°
Koba, Bangka, Indonesia	+7	06.21	07.23	08.35	8^m 49^s	20°
Bakong, Bangka , Indonesia	+7	06.21	07.24	08.35	2^m 05^s	20°
Tanjungpandan, Belitung, Indon'	+7	06.21	07.24	08.36	1^m 14^s	21°
Manggar, Belitung, Indonesia	+7	06.21	07.25	08.37	2^m 07^s	22°
Kendawangan, Borneo, Indon'	+7	06.22	07.27	08.40	2^m 20^s	24°
Pangkalanbuun, Borneo, Indon'	+7	06.23	07.28	08.43	2^m 08^s	26°
Sampit, Borneo, Indonesia	+7	06.23	07.29	08.45	2^m 49^s	28°
Palangkaraya, Borneo, Indonesia	+7	06.23	07.30	08.47	2^m 31^s	29°
Tanjung, Borneo, Indonesia	+8	07.24	08.32	09.50	2^m 33^s	31°
Tanahgrogot, Borneo, Indonesia	+8	07.25	08.34	09.52	2^m 35^s	32°
Kulawi, Sulawesi, Indonesia	+8	07.28	08.39	10.01	2^m 51^s	38°
Poso, Sulawesi, Indonesia	+8	07.28	08.40	10.02	2^m 47^s	38°
Ampana, Sulawesi, Indonesia	+8	07.30	08.42	10.05	2^m 51^s	39°
Tataba, Banggai, Indonesia	+8	07.30	08.43	10.07	1^m 50^s	41°
Ternate, Halmahera, Indonesia	+9	08.36	09.53	11.21	2^m 25^s	47°
Weda, Halmahera, Indonesia	+9	08.37	09.54	11.22	3^m 15^s	49°
Wayamli, Halmahera, Indonesia	+9	08.38	09.56	11.25	2^m 58^s	50°

2016 September 1, annular

Annular: Gabon, Congos, Tanzania, northern Mozambique, Madagascar, Réunion
Partial: Africa, Indian Ocean

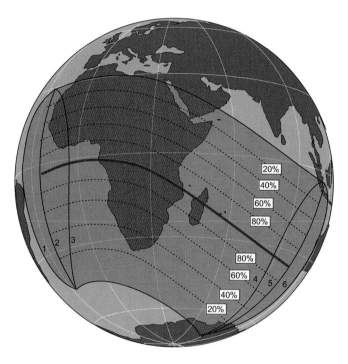

Course of the eclipse

1 End of eclipse at sunrise

2 Maximum at sunrise

3 Beginning of eclipse at sunrise

4 End of eclipse at sunset

5 Maximum at sunset

6 Beginning of eclipse at sunset

Three years after the eclipse of 2013, another one traverses central Africa. The annular solar eclipse, in which there will be maximum cover of 95%, begins in the middle of the Atlantic. Its 120 km-wide path reaches Gabon at Omboué, a mere 110 km further

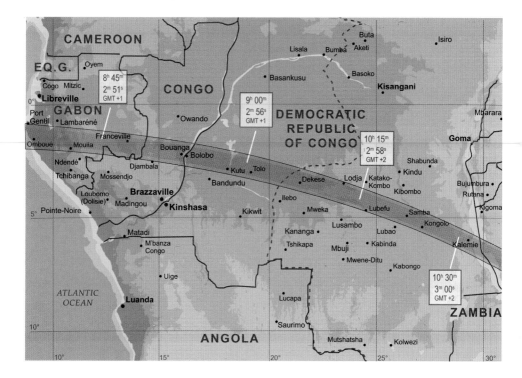

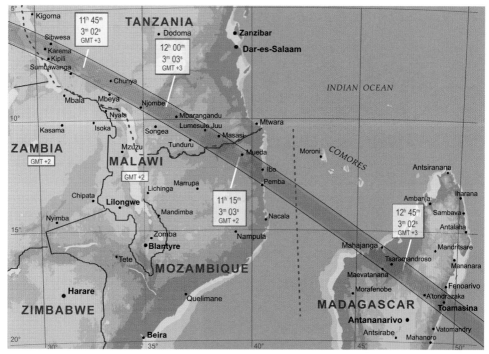

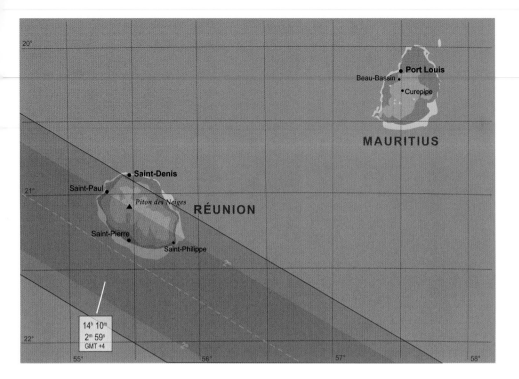

south than in 2013. The eclipse then sweeps across the highlands of Gabon before crossing the Congo River at Bouanga and Bolobo, and is 100 km wide when it reaches Lake Tanganyika. In Tanzania it crosses the main road from the coast to Zambia north-east of Mbeya. The umbra leaves the mainland at the coast of northern Mozambique, reaching Madagascar at Mahajanga. From there it crosses the island passing through the east-coast city of Toamasina for 2 minutes 40 seconds before it moves out into the Indian Ocean.

The island of Réunion, which falls under its central track east of Madagascar, is undoubtedly an interesting spot from which to observe the eclipse. The French island's peak, Piton des Neiges, is over 3000 m high. That location, above all the west side, will provide an impressive view of the approaching eclipse. Here the shadow has a speed of 2700 km/h. From the island of Mauritius, 140 km from the centre of the eclipse, coverage of only 90%. The eclipse ends 1350 km off the Australian coast.

Weather

At the time of the eclipse Gabon is still in the dry season. Rainfall does not start until October. Yet there is a tropical climate with high ambient humidity and frequent cloudy skies. It is also still the dry season in Congo and Tanzania; here the degree of cloud cover decreases towards the centre of the continent with a high probability of broken cloud. The same applies to Tanzania. Towards the east coast the probability of cloud rises again to about 40%.

Madagascar has a sub-tropical climate. This eclipse fortunately occurs in the dry season because the monsoon-like rains only start in December. The weather conditions on the west side of the island are somewhat better than in the east.

2016 Sep 1, Thursday Location	Time zone (GMT ±)	Begin- ning	Maxi- mum	End	Dura- tion	Height of Sun	Cover- age
Omboué, Gabon	+1	07.19	08.40	10.18	2m 42s	34°	94%
Mouila, Gabon	+1	07.20	08.43	10.24	2m 11s	36°	94%
Franceville, Gabon	+1	07.23	08.48	10.32	2m 39s	40°	94%
Bouanga / Bolobo, Congo / DRC	+1	07.25	08.53	10.40	2m 52s	43°	94%
Kutu, Dem. Rep. Congo	+1	07.28	08.58	10.47	2m 51s	46°	94%
Lodja, Dem. Rep. Congo	+2	08.36	10.13	12.05	1m 26s	55°	95%
Kongolo, Dem. Rep. Congo	+2	08.44	10.25	12.20	2m 46s	60°	95%
Kalemie, Dem. Rep. Congo	+2	08.48	10.32	12.28	2m 52s	63°	95%
Karema, Tanzania	+3	09.52	11.38	13.33	2m 55s	65°	95%
Chunya, Tanzania	+3	10.01	11.50	13.45	2m 08s	68°	95%
Njombe, Tanzania	+3	10.06	11.56	13.50	1m 33s	69°	95%
Mbarangandu, Tanzania	+3	10.12	12.03	13.57	2m 53s	70°	95%
Masasi, Tanzania	+3	10.18	12.11	14.02	0m 23s	70°	95%
Mueda, Mozambique	+2	09.22	11.15	13.06	3m 01s	70°	95%
Ibo, Mozambique	+2	09.26	11.19	13.09	3m 02s	70°	95%
Mahajanga, Madagascar	+3	10.48	12.41	14.25	2m 12s	64°	95%
A'tondrazaka, Madagascar	+3	10.58	12.50	14.30	1m 20s	60°	95%
Toamasima, Madagascar	+3	11.01	12.53	14.32	2m 42s	59°	95%
Saint-Paul, Réunion	+4	12.22	14.09	15.42	2m 13s	50°	94%
Saint-Pierre, Réunion	+4	12.24	14.10	15.43	2m 53s	50°	94%

2017 February 26, annular

Annular: Chile, Argentina, Angola
Partial: Southern South America, western Africa

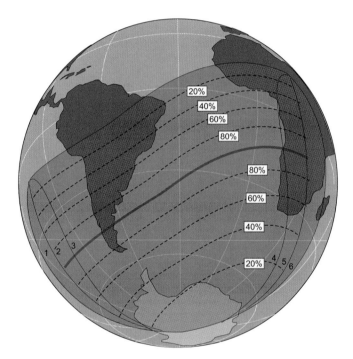

Course of the eclipse

1 *End of eclipse at sunrise*

2 *Maximum at sunrise*

3 *Beginning of eclipse at sunrise*

4 *End of eclipse at sunset*

5 *Maximum at sunset*

6 *Beginning of eclipse at sunset*

This annular solar eclipse begins 3100 km off the Chilean coast. The 56 km-wide band of shadow reaches the broken coast of southern Chile north of the Laguna San Rafael National Park. The towns of Puerto Aisén and Puerto Chacabuco are the first habitations to be reached by the annular eclipse. At the time of the central shadow, the Sun has already risen to 30° above the

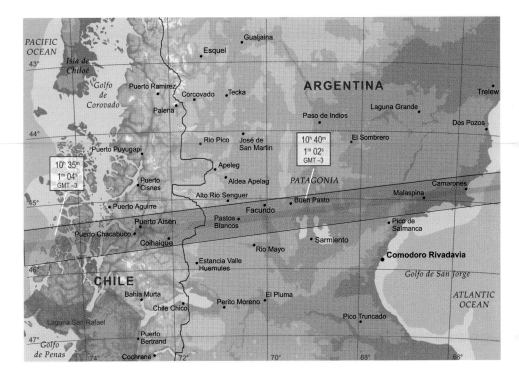

horizon. The 2.5% of the Sun's surface which remains uncovered will form an enchanting ring of light over the landscape. The celestial spectacle can be observed from Coyhaique, high in the Andes. The narrow 53 km-wide band of shadow moves on to Argentina, passing the town of Buen Pasto as well as Musters and Colhué Lakes which are located in the thinly populated shadow zone. The shadow leaves the South American continent at Camarones, north of Comodoro Rivadavia and crosses the South Atlantic in a north-easterly direction.

In the South Atlantic, midway along the path of the eclipse, it has narrowed to 30 km. At the same time less than 1% of the Sun remains uncovered. Such a thin ring will produce a breath-taking view. It is so thin that the uneven Moon landscape will break it up into individual fragments which will then sparkle on the edge of the Moon like pearls of light.

In Angola it is the coastal town of Lucira which welcomes the eclipse on African soil. Its path has grown back to a width of

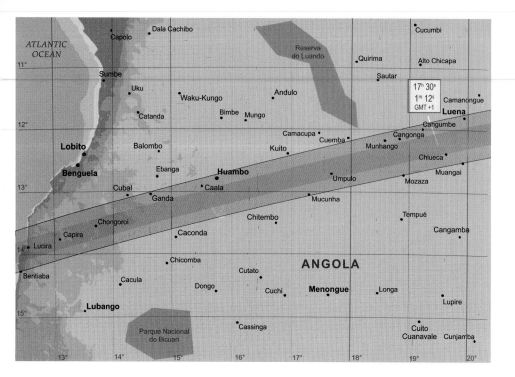

ATLANTIC OCEAN

11°
Capolo • Dala Cachibo
• Cucumbi

Reserva do Luando
• Quirima
• Alto Chicapa

Sumbe
• Sautar

17h 30s
1m 12s
GMT +1

• Uku
Andulo
• Camanongue

• Waku-Kungo
Bimbe • Mungo
Luena

12°
• Catanda
Camacupa •
Cuemba •
Cangonga
• Cangumbe

Lobito
Balombo •
Kuito
Munhango
Chiueca

Ebanga
Kuito
• Umpulo
Mozaza
Muangai

Benguela
• Cubal
Huambo
• Caala

13°
Ganda
• Mucunha

Chongoroi
Chitembo
Tempué
• Cangamba

14°
Capira
Caconda

• Lucira
Chicomba
ANGOLA

Bentiaba
Cutato

Cacula
Dongo •
Cuchi
Menongue
Longa
Lupire

Lubango

15°
Parque Nacional do Bicuar
• Cassinga
Cuito Cuanavale
Cunjamba

13° 14° 15° 16° 17° 18° 19° 20°

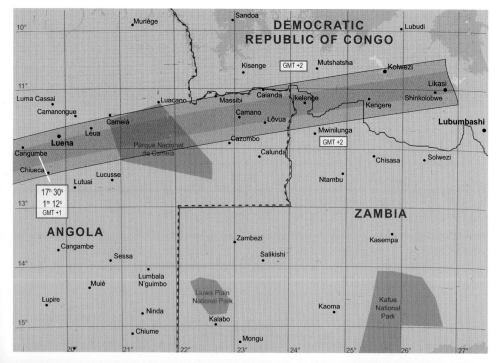

10°
• Muriége
Sandoa
DEMOCRATIC REPUBLIC OF CONGO
• Lubudi

Kisenge
GMT +2
Mutshatsha
Kolwezi

11°
Luma Cassai
Luacano
Massibi
Caianda
Ikelenge
Shinkolobwe
Likasi

Camanongue
Cameiá
Camano
Lôvua
Kengere

Léua
Cazombo
Mwinilunga
Lubumbashi

Luena
Calunda
GMT +2

Cangumbe
Lucusse
Chisasa
Solwezi

Chiueca

17h 30s
1m 12s
GMT +1
Lutuai
Ntambu

12°

13°
ANGOLA
ZAMBIA

Cangambe
Zambezi
Kasempa

Sessa
Salikishi

14°
Lumbala N'guimbo

Muié
Liuwa Plain National Park
Kaoma
Kafue National Park

Lupire
Ninda

15°
Chiume
Kalabo
Mongo

20° 21° 22° 23° 24° 25° 26° 27°

71 km. It subsequently travels into the highlands of Angola which in the south of the country are more like savannah. It crosses the north-west tip of Zambia where the source of the mighty Zambesi River lies. The political situation at the time may well determine which of the three countries is most easily accessible. Not far from Lubumbashi at the southern tip of the Democratic Republic of Congo the shadow leaves the Earth. The end of an annular solar eclipse is an impressive sight: surrounded by the colours of dusk, a red ringed Sun thrones on the horizon above the landscape.

2017 Feb 26, Sunday Location	Time zone (GMT ±)	Begin-ning	Maxi-mum	End	Dura-tion	Height of Sun	Cover-age
Puerto Aisén, Chile	−3	09.23	10.36	11.56	0^m 58^s	33°	97%
Coihaique, Chile	−3	09.24	10.37	11.57	0^m 55^s	33°	97%
Pastos Blancos, Argentina	−3	09.24	10.38	11.59	1^m 02^s	34°	97%
Buen Pasto, Argentina	−3	09.25	10.40	12.01	0^m 55^s	35°	95%
Malaspina, Argentina	−3	09.27	10.43	12.05	0^m 55^s	37°	98%
Camarones, Argentina	−3	09.27	10.44	12.07	0^m 56^s	37°	98%
Lucira, Angola	+1	16.16	17.27	18.29	1^m 07^s	16°	96%
Capira, Angola	+1	16.17	17.27	18.29	1^m 07^s	14°	96%
Chongoroi, Angola	+1	16.18	17.28	18.26	1^m 08^s	13°	96%
Ganda, Angola	+1	16.19	17.28	18.22	0^m 20^s	13°	96%
Huambo, Angola	+1	16.21	17.29	18.18	0^m 25^s	12°	96%
Umpulo, Angola	+1	16.22	17.30	18.10	1^m 09^s	10°	96%
Cangonga, Angola	+1	16.24	17.30	18.05	0^m 52^s	9°	96%
Luena, Angola	+1	16.25	17.30	18.00	0^m 09^s	7°	96%
Lôvua, Angola	+1	16.27	17.31	17.46	1^m 12^s	3°	96%
Ikelenge, Zambia	+2	17.28	18.31	18.43	1^m 12^s	2°	96%
Kengere, Dem. Rep. Congo	+2	17.28	18.31	18.38	1^m 15^s	2°	96%
Likasi, Dem. Rep. Congo	+2	17.29	18.32	18.32	0^m 30^s	1°	95%

2017 August 21, total

Total: United States of America, Atlantic
Partial: North, Central and northern South America, West coast
of Europe and North Africa

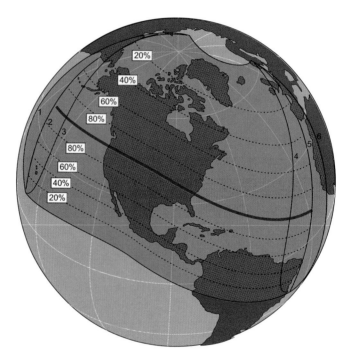

Course of the eclipse

1 *End of eclipse at sunrise*

2 *Maximum at sunrise*

3 *Beginning of eclipse at sunrise*

4 *End of eclipse at sunset*

5 *Maximum at sunset*

6 *Beginning of eclipse at sunset*

Almost 4000 km off the American west coast, the eclipse of August 21, begins in the North Pacific. It belongs to the same Saros cycle as the last central eclipse of the twentieth century on August 11, 1999, which crossed Europe and the Middle East. A

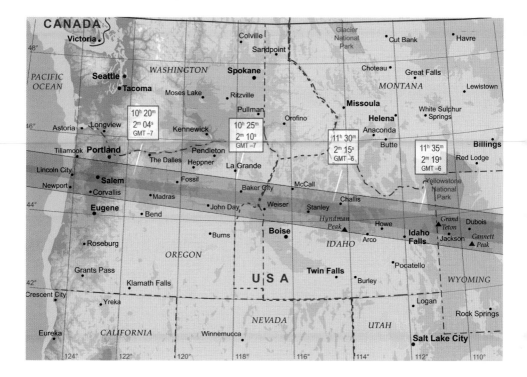

total solar eclipse crosses the United States again for the first time in 38 years (the eclipse of 2012 was annular).

The band of shadow reaches Oregon at 9.05 local (Pacific Daylight Saving) time. The Sun has already reached a height of 27° above the horizon when the Moon starts to edge in front of it. One hour later, at 10.17 and with the Sun having risen to 39°, the rapid approach of the eclipse from the Pacific can be observed from the coastal cities of Lincoln City or Newport. One minute later the first larger city, Salem, the capital of Oregon, is immersed in darkness. The season could hardly be better for weather. On the west coast and in the centre of the continent late summer is the period of least rain and cloud. This applies particularly east of the Cascade mountains. This mountain range forms a meteorological divide between the humid coastal climate and the dry conditions inland.

Once the eclipse has crossed the forests of Oregon, its shadow reaches the Rocky Mountains in Idaho. Take care with the

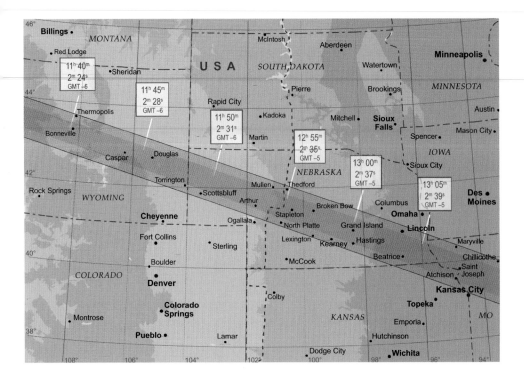

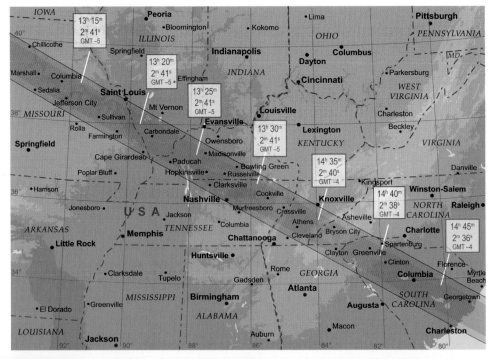

boundary between Pacific and Mountain Time which does not follow the state line here. Depending on the weather, cloud can be formed here through rising air masses. The shadow zone passes close by Boise, the capital of Idaho, and then over the 4200 m (13 770 ft) high Grand Teton just south of Yellowstone National Park. This largest and oldest of American nature reserves, with almost 3000 geysers and an elemental landscape, offers a force of nature to equal that of the solar eclipse.

The duration of the eclipse steadily increases as it travels across the American continent. Dubois, Wyoming, at the foot of Gannet Peak is already covered in darkness for 2 mintues 20 seconds. The lunar shadow here races across the eastern Rocky Mountains at a 'mere' 2900 km/h to reach the Great Plains, the dry, fertile prairies at the centre of the continent. The shadow is now 100 km (60 miles) wide. In Nebraska again watch for the change from Mountain to Central Time which does not follow the state line.

Passing just north of Kansas City, for a time it follows the Missouri and crosses the Mississippi just south of Saint Louis. The centre of the eclipse is located north of Nashville, Tennessee. The city itself is only subject to 1 minute 30 seconds of darkness, however, totality grows to 2 mintues 40 seconds only 80 km (50 miles) further north, where the smaller towns such as in Hopkinsville and Russelville, Kentucky, and Cookville, Tennessee, are situated. The shadow has slowed down the most here to 2300 km/h (1400 mph). At this location the Sun has risen to 65° above the horizon at the time of maximum, giving the observer the impression that the Sun is almost in the zenith.

The eclipse passes through Great Smoky National Park between Tennessee and North Carolina where it crosses the Appalachains. However, the probability of a cloudy sky increases in comparison to the plains.

At 14.48 (2.48 pm) the shadow leaves the American continent at Georgetown, South Carolina. The shadow continues to track across the Atlantic Ocean, finally leaving the Earth 1000 km off the West African coast. Seven years later, on April 8, 2024, is the next occasion when an eclipse travels across the USA.

2017 Aug 21, Monday Location	Time zone (GMT ±)	Begin- ning	Maxi- mum	End	Dura- tion	Height of Sun
Lincoln City, Oregon, USA	−7	09.04	10.17	11.36	1m 58s	39°
Salem, Oregon, USA	−7	09.05	10.18	11.38	1m 59s	40°
Corvallis, Oregon, USA	−7	09.05	10.18	11.38	1m 30s	40°
Madras, Oregon, USA	−7	09.07	10.21	11.41	2m 00s	42°
John Day, Oregon, USA	−7	09.09	10.24	11.45	1m 54s	44°
Weiser, Idaho, USA	−6	10.10	11.27	12.49	2m 05s	45°
Stanley, Idaho, USA	−6	10.12	11.30	12.53	2m 15s	47°
Howe, Idaho, USA	−6	10.14	11.33	12.56	2m 03s	49°
Idaho Falls, Idaho, USA	−6	10.15	11.34	12.58	1m 32s	50°
Jackson, Wyoming, USA	−6	10.17	11.36	13.01	2m 10s	51°
Dubois, Wyoming, USA	−6	10.18	11.38	13.03	2m 21s	51°
Bonneville, Wyoming, USA	−6	10.20	11.41	13.06	2m 24s	53°
Casper, Wyoming, USA	−6	10.22	11.44	13.10	2m 27s	54°
Douglas, Wyoming, USA	−6	10.24	11.46	13.12	2m 27s	55°
Scottsbluff, Nebraska, USA	−6	10.26	11.49	13.16	1m 15s	56°
Arthur, Nebraska, USA	−6	10.29	11.53	13.20	2m 14s	57°
Mullen, Nebraska, USA	−6	10.30	11.54	13.21	1m 48s	58°
North Platte, Nebraska, USA	−5	11.30	12.55	14.22	1m 26s	58°
Stapleton, Nebraska, USA	−5	11.31	12.56	14.22	2m 34s	59°
Kearney, Nebraska, USA	−5	11.33	12.59	14.26	1m 39s	60°
Grand Island, Nebraska, USA	−5	11.34	13.00	14.27	2m 36s	61°
Hasting, Nebraska, USA	−5	11.34	13.00	14.27	2m 01s	61°
Lincoln, Nebraska, USA	−5	11.37	13.03	14.30	1m 40s	61°
Beatrice, Nebraska, USA	−5	11.37	13.04	14.31	2m 31s	61°
Maryville, Missouri, USA	−5	11.41	13.04	14.34	0m 47s	61°
Atchison, Kansas, USA	−5	11.40	13.08	14.35	2m 08s	61°
St Joseph, Missouri, USA	−5	11.41	13.08	14.35	2m 39s	61°
Marshall, Missouri, USA	−5	11.44	13.12	14.39	2m 38s	63°
Columbia, Missouri, USA	−5	11.46	13.14	14.41	2m 40s	63°
Jefferson City, Missouri, USA	−5	11.46	13.15	14.41	2m 22s	63°
Sullivan, Missouri, USA	−5	11.48	13.17	14.44	2m 26s	64°
St Louis, Missouri, USA	−5	11.51	13.18	14.45	0m 57s	64°
Farmington, Missouri, USA	−5	11.50	13.19	14.46	1m 38s	64°
Cape Girardeau, Missouri, USA	−5	11.52	13.21	14.48	1m 28s	64°
Carbondale, Illinois, USA	−5	11.53	13.22	14.49	2m 40s	64°

2017 Aug 21, Monday Location	Time zone (GMT ±)	Begin- ning	Maxi- mum	End	Dura- tion	Height of Sun
Paducah, Kentucky, USA	−5	11.54	13.24	14.50	2ᵐ 08ˢ	64°
Hopkinsville, Kentucky, USA	−5	11.57	13.26	14.52	2ᵐ 40ˢ	64°
Clarksville, Tennessee, USA	−5	11.57	13.27	14.53	2ᵐ 08ˢ	64°
Russellville, Kentucky, USA	−5	11.58	13.28	14.53	2ᵐ 35ˢ	63°
Bowling Green, Kentucky, USA	−5	11.59	13.28	14.54	1ᵐ 22ˢ	63°
Nashville, Tennessee, USA	−5	11.59	13.29	14.54	1ᵐ 43ˢ	63°
Cookville, Tennessee, USA	−5	12.01	13.31	14.56	2ᵐ 39ˢ	64°
Crossville, Tennessee, USA	−5	12.03	13.32	14.57	2ᵐ 37ˢ	64°
Athens, Tennessee, USA	−4	13.04	14.34	15.59	2ᵐ 32ˢ	63°
Bryson City, N Carolina, USA	−4	13.07	14.36	16.01	2ᵐ 09ˢ	63°
Clayton, Georgia, USA	−4	13.07	14.37	16.02	2ᵐ 31ˢ	63°
Greenville, South Carolina, USA	−4	13.10	14.39	16.04	2ᵐ 20ˢ	63°
Spartanburg, S Carolina, USA	−4	13.10	14.40	16.04	0ᵐ 28ˢ	63°
Clinton, South Carolina, USA	−4	13.11	14.41	16.05	2ᵐ 32ˢ	62°
Columbia, South Carolina, USA	−4	13.13	14.43	16.07	2ᵐ 32ˢ	62°
Charleston, South Carolina, USA	−4	13.17	14.47	16.10	1ᵐ 21ˢ	62°
Georgetown, S Carolina, USA	−4	13.17	14.48	16.10	1ᵐ 59ˢ	62°
Lincoln City, Oregon, USA	−7	09.04	10.17	11.36	1ᵐ 58ˢ	39°
Salem, Oregon, USA	−7	09.05	10.18	11.38	1ᵐ 59ˢ	40°
Corvallis, Oregon, USA	−7	09.05	10.18	11.38	1ᵐ 30ˢ	40°
Madras, Oregon, USA	−7	09.07	10.21	11.41	2ᵐ 00ˢ	42°
John Day, Oregon, USA	−7	09.09	10.24	11.45	1ᵐ 54ˢ	44°
Weiser, Idaho, USA	−6	10.10	11.27	12.49	2ᵐ 05ˢ	45°
Stanley, Idaho, USA	−6	10.12	11.30	12.53	2ᵐ 15ˢ	47°
Howe, Idaho, USA	−6	10.14	11.33	12.56	2ᵐ 03ˢ	49°
Idaho Falls, Idaho, USA	−6	10.15	11.34	12.58	1ᵐ 32ˢ	50°
Jackson, Wyoming, USA	−6	10.17	11.36	13.01	2ᵐ 10ˢ	51°
Dubois, Wyoming, USA	−6	10.18	11.38	13.03	2ᵐ 21ˢ	51°
Bonneville, Wyoming, USA	−6	10.20	11.41	13.06	2ᵐ 24ˢ	53°
Casper, Wyoming, USA	−6	10.22	11.44	13.10	2ᵐ 27ˢ	54°
Douglas, Wyoming, USA	−6	10.24	11.46	13.12	2ᵐ 27ˢ	55°
Scottsbluff, Nebraska, USA	−6	10.26	11.49	13.16	1ᵐ 15ˢ	56°
Arthur, Nebraska, USA	−6	10.29	11.53	13.20	2ᵐ 14ˢ	57°
Mullen, Nebraska, USA	−6	10.30	11.54	13.21	1ᵐ 48ˢ	58°
North Platte, Nebraska, USA	−5	11.30	12.55	14.22	1ᵐ 26ˢ	58°

2017 Aug 21, Monday Location	Time zone (GMT ±)	Begin- ning	Maxi- mum	End	Dura- tion	Height of Sun
Stapleton, Nebraska, USA	−5	11.31	12.56	14.22	2ᵐ 34ˢ	59°
Kearney, Nebraska, USA	−5	11.33	12.59	14.26	1ᵐ 39ˢ	60°
Grand Island, Nebraska, USA	−5	11.34	13.00	14.27	2ᵐ 36ˢ	61°
Hasting, Nebraska, USA	−5	11.34	13.00	14.27	2ᵐ 01ˢ	61°
Lincoln, Nebraska, USA	−5	11.37	13.03	14.30	1ᵐ 40ˢ	61°
Beatrice, Nebraska, USA	−5	11.37	13.04	14.31	2ᵐ 31ˢ	61°
Maryville, Missouri, USA	−5	11.41	13.04	14.34	0ᵐ 47ˢ	61°
Atchison, Kansas, USA	−5	11.40	13.08	14.35	2ᵐ 08ˢ	61°
St Joseph, Missouri, USA	−5	11.41	13.08	14.35	2ᵐ 39ˢ	61°
Marshall, Missouri, USA	−5	11.44	13.12	14.39	2ᵐ 38ˢ	63°
Columbia, Missouri, USA	−5	11.46	13.14	14.41	2ᵐ 40ˢ	63°
Jefferson City, Missouri, USA	−5	11.46	13.15	14.41	2ᵐ 22ˢ	63°
Sullivan, Missouri, USA	−5	11.48	13.17	14.44	2ᵐ 26ˢ	64°
St Louis, Missouri, USA	−5	11.51	13.18	14.45	0ᵐ 57ˢ	64°
Farmington, Missouri, USA	−5	11.50	13.19	14.46	1ᵐ 38ˢ	64°
Cape Girardeau, Missouri, USA	−5	11.52	13.21	14.48	1ᵐ 28ˢ	64°
Carbondale, Illinois, USA	−5	11.53	13.22	14.49	2ᵐ 40ˢ	64°
Paducah, Kentucky, USA	−5	11.54	13.24	14.50	2ᵐ 08ˢ	64°
Hopkinsville, Kentucky, USA	−5	11.57	13.26	14.52	2ᵐ 40ˢ	64°
Clarksville, Tennessee, USA	−5	11.57	13.27	14.53	2ᵐ 08ˢ	64°
Russellville, Kentucky, USA	−5	11.58	13.28	14.53	2ᵐ 35ˢ	63°
Bowling Green, Kentucky, USA	−5	11.59	13.28	14.54	1ᵐ 22ˢ	63°
Nashville, Tennessee, USA	−5	11.59	13.29	14.54	1ᵐ 43ˢ	63°
Cookville, Tennessee, USA	−5	12.01	13.31	14.56	2ᵐ 39ˢ	64°
Crossville, Tennessee, USA	−5	12.03	13.32	14.57	2ᵐ 37ˢ	64°
Athens, Tennessee, USA	−4	13.04	14.34	15.59	2ᵐ 32ˢ	63°
Bryson City, N Carolina, USA	−4	13.07	14.36	16.01	2ᵐ 09ˢ	63°
Clayton, Georgia, USA	−4	13.07	14.37	16.02	2ᵐ 31ˢ	63°
Greenville, South Carolina, USA	−4	13.10	14.39	16.04	2ᵐ 20ˢ	63°
Spartanburg, S Carolina, USA	−4	13.10	14.40	16.04	0ᵐ 28ˢ	63°
Clinton, South Carolina, USA	−4	13.11	14.41	16.05	2ᵐ 32ˢ	62°
Columbia, South Carolina, USA	−4	13.13	14.43	16.07	2ᵐ 32ˢ	62°
Charleston, South Carolina, USA	−4	13.17	14.47	16.10	1ᵐ 21ˢ	62°
Georgetown, S Carolina, USA	−4	13.17	14.48	16.10	1ᵐ 59ˢ	62°

Special Accounts
of Solar Eclipses

The Eclipse of the Sun, 1820

WILLIAM WORDSWORTH

High on her speculative tower
Stood Science waiting for the hour
When Sol was destined to endure
'That' darkening of his radiant face
Which Superstition strove to chase,
Erewhile, with rites impure.

Afloat beneath Italian skies,
Through regions fair as Paradise
We gaily passed, — till Nature wrought
A silent and unlooked-for change,
That checked the desultory range
Of joy and sprightly thought.

Where'er was dipped the toiling oar,
The waves danced round us as before,
As lightly, though of altered hue,
'Mid recent coolness, such as falls
At noontide from umbrageous walls
That screen the morning dew.

No vapour stretched its wings; no cloud
Cast far or near a murky shroud;

The sky an azure field displayed;
'Twas sunlight sheathed and gently charmed,
Of all its sparkling rays disarmed,
And as in slumber laid, —

Or something night and day between,
Like moonshine — but the hue was green;
Still moonshine, without shadow, spread
On jutting rock, and curved shore,
Where gazed the peasant from his door
And on the mountain's head.

It tinged the Julian steeps — it lay,
Lugano! on thy ample bay;
The solemnizing veil was drawn
O'er villas, terraces, and towers;
To Albogasio's olive bowers,
Porlezza's verdant lawn.

But Fancy with the speed of fire
Hath passed to Milan's loftiest spire,
And there alights 'mid that aerial host
Of Figures human and divine,
White as the snows of Apennine
Indurated by frost.

Awe-stricken she beholds the array
That guards the Temple night and day;
Angels she sees — that might from heaven have flown,
And Virgin-saints, who not in vain
Have striven by purity to gain
The beatific crown —

Sees long-drawn files, concentric rings
Each narrowing above each; — the wings,
The uplifted palms, the silent marble lips
The starry zone of sovereign height —

All steeped in this portentous light!
All suffering dim eclipse!

Thus after Man had fallen (if aught
These perishable spheres have wrought
May with that issue be compared)
Throngs of celestial visages,
Darkening like water in the breeze,
A holy sadness shared.

Lo! while I speak, the labouring Sun
His glad deliverance has begun:
The cypress waves her sombre plume
More cheerily; and town and tower,
The vineyard and the olive-bower,
Their lustre re-assume!

O Ye, who guard and grace my home
While in far-distant lands we roam,
What countenance hath this Day put on for you?
While we looked round with favoured eyes,
Did sullen mists hide lake and skies
And mountains from your view?

Or was it given you to behold
Like vision, pensive though not cold,
From the smooth breast of gay Winandermere?
Saw ye the soft yet awful veil
Spread over Grasmere's lovely dale,
Helvellyn's brow severe?

I ask in vain — and know far less
If sickness, sorrow, or distress
Have spared my Dwelling to this hour;
Sad blindness! but ordained to prove
Our faith in Heaven's unfailing love
And all-controlling power.

The eclipse of July 8, 1842

ADALBERT STIFTER

There are things we have known for fifty years, and suddenly, when we reach fifty-one, we are amazed by the weight and fruitfulness of their content. That is how I felt about the total solar eclipse that I experienced in the early hours of a clear Vienna morning on July 8, 1842. I am quite able to set out such an event on paper by a sketch and calculations; and I knew that at a certain hour the Moon would coincide with the path of the Sun and the Earth would cut off a section of its conical shadow, which would then draw a black line across the globe due to the continued progress of the Moon in its trajectory and the Earth spinning around its axis — something which is seen at various locations as a disk appearing to cover the Sun, taking more and more of it away until nothing is left but a small sickle, which also disappears in the end. On Earth it grows darker and darker, until the Sun's sickle appears again on the other side, and grows until gradually full daylight is restored. All of this I knew in advance, and knew it so well that I believed myself able to describe a full solar eclipse in advance as though I had already seen it. But in the course of the event, as I was standing at a spot high above the whole city and saw the phenomenon with my own eyes, completely different things happened, of course, of which I had no inkling even in my wildest dreams, and which no one thinks of if they have not seen such a miracle.

Never in my whole life was I so shaken as in these two minutes; it was as though God had suddenly spoken clearly and I had understood. I came down from my observation spot like Moses might have done thousands of years ago from the fiery mountain: confused and with a frozen heart.

It was such a simple thing. One body shines on another one and the latter casts its shadow on a third body: but these bodies are so far apart that we have no way of imagining it, they are so gigantic that they extend far beyond anything we call big — such a complex of phenomena is associated with this simple event, this physical process is endowed with such a moral power, that it mounts up in our heart into an incomprehensible miracle.

A thousand times a thousand years ago, God created the conditions for an eclipse to occur today at this second; and He laid the seeds in our heart so that we could experience and feel it. Into the script of His stars he laid the promise that it would happen after thousands and thousands of years, and our fathers learnt to decipher the script and predict the second in which it would occur; we, their grandchildren, direct our eyes and telescopes at the preordained moment towards the Sun and behold: there it is. Reason has triumphed in that it has managed to learn from Him and calculate the magnificence and composition of His heavens — and, indeed, this is a righteous triumph for human beings to claim. It arrives, and quietly it continues to grow, and we become aware that God also gave human beings something for their heart which we did not know in advance, and which is a millionfold greater in value than what we have learned and can calculate in advance by reason: He gave them the Word: 'I am — not because these bodies and these phenomena exist, no, I am because that is what your hearts tell you in awe at this moment, and because those hearts experiences their own greatness because of their awe' — animals fear, humans worship. ...

At 5 o'clock I climbed up to the observation spot of house number 495 in the city, from where there is a view not just of the whole city, but also of the surrounding countryside to the farthest horizon, where the Hungarian mountains shimmer in the dawn like delicate mirages. The Sun had already risen and its friendly light shone on the steaming Danube meadows, the reflecting waters and the angular forms of the city — particularly St Stephen's Cathedral which rose out of the city like a dark, quiet mountain range so close that one could almost touch it.

I looked at the Sun, which was to be the subject of such strange events in a few minutes, with a peculiar feeling. In the distance, where the great river lies, there was a thick, extended line of mist; clusters of fog and cloud also crept around on the south-eastern horizon, which we feared greatly, and whole districts of the city were suspended in the haze. There were only very thin veils where the Sun stood and they too revealed large blue islands.

The instruments were set up, the glass for viewing the Sun prepared, but the time had not yet come. Below, the rattling of the carriages, the hustle and bustle began — above, people wanting to watch the eclipse gathered; our observation spot began to fill, heads were looking out of the dormer windows of surrounding houses, figures were standing on roof ridges all looking at the same spot in the sky; there was even a group at the highest tip of the tower of St Stephen's Cathedral, on the very highest platform of the scaffolding, just as trees often find a niche on a rock in which they manage to grow. Thousands of eyes were probably looking at the Sun from the surrounding mountains, the same Sun that for millennia had cast its blessings on the Earth without a word of thanks from anyone — today it was the target of millions of eyes, but observed through smoky glass it continued to hover as a red or green sphere in space, pure and beautifully rounded.

Finally, at the predicted minute, as if an invisible angel had given it a gentle kiss of death, a fine band of the sun's light began to retreat from the breath cast by this kiss while the other side continued to well gently and golden in the lens of the telescope. 'It's coming' the call went up also among those who had observed the Sun only through blackened glass, but otherwise with bare eyes — 'it's coming,' and everyone watched in excitement wondering what would happen next.

The first strange, alien feeling now began to seep into our hearts, namely that out there thousands and millions of miles distant, where human beings have never penetrated, something was happening on Earth, at the long predetermined moment, to bodies with whose nature no human being was familiar.

People might suggest that the whole event is quite natural and easy to calculate from the laws governing the motion of these bodies; the wonderful magic of the beauty which God gave these things does not bother with such calculations. It exists because it exists, indeed, it exists in spite of such calculations and blessed is the heart which can experience it. Because that alone is wealth, there is no other — the sublime which overwhelms our soul lives in the immense space of the heavens and yet mathematics considers this space to have no quality other than its size.

While everyone was looking, moving a telescope here and a telescope there, drawing each other's attention to various things that were happening, the invisible darkness increasingly encroached on the beautiful light of the Sun — everyone was full of anticipation, the excitement rose; but so mighty and full was the ocean of light showering down from the Sun that there was no feeling of privation; the clouds continued to shine, the strip of water shimmered, the birds darted across the roofs, the towers of St Stephen's Cathedral cast their peaceful shadows on the sparkling roof, people were driving and riding across the bridge as usual, little knowing that the balsam of life, the light, was secretly ebbing away. Outside, however, at the Kahlengebirge mountain and beyond the Belvedere Palace, the darkness — or rather a leaden light — was stealing closer like a wild animal. And yet it might have been an illusion, our observation spot remained lively and bright and the cheeks and faces of those nearby were as clear and friendly as always.

The strange thing was that this eerie mass, this profoundly black advancing entity which was slowly eating away the Sun should be our Moon, the beautiful gentle Moon which on other occasions cast its silvery light in the night; and yet that is what it was and in the telescope its edges set with spikes and bulges also appeared, the terrible mountains piling up on the sphere, smile so sweetly at us.

Finally, the effect also became visible on Earth — more and more so as the glowing sickle in the sky became smaller and smaller; the river, no longer shimmering, became a ribbon of grey taffeta, matt shadows lay everywhere, the swallows became restless, the beautifully gentle radiance of the sky was extinguished as if frosted by a breath; we felt a cool breeze arise, an indescribably strange but leaden light cast its spectre across the meadows; across the forests their gentle movement disappeared with the play of the light and peace rested on them, not the peace of sleep, however, but of impotence. The light over the landscape turned more and more sallow, and the landscape itself became more and more rigid — our shadows were cast empty and tenuous upon the walls, our faces became ashen. This gradual decay in the midst of what a few minutes ago had been a fresh morning, was dreadful.

We had imagined the gradual disappearance of the light rather like the failing of the evening light only without the redness of the evening sky; we had never imagined how eerie an onset of evening without the redness of the evening sky could be. But in other respects, too, this was a quite different twilight, it was a burdensome, uncanny alienation of our nature. Towards the south-east there was an alien, yellowish reddish shadow and the mountains and even the Belvedere were submerged in it; at our feet the city sank deeper and deeper, a shadow play without being, the driving, walking and riding across the bridge occurred as though we saw it in a black mirror. As the tension rose to its highest degree, I took one last look through the telescope, the very last one: thin, like an incision with a knife in the darkness, the glowing sickle stood there, about to be extinguished at any moment, and as I lifted my unprotected eyes I saw that everyone else had also put away their telescopes and was looking up with bare eyes. They no longer needed them because, in the same way as the last spark of an extinguished wick disappears, the last spark of the Sun melted away, probably through the gap between two lunar mountains. It was an exceedingly sad moment as the one disc covered the other, and in actual fact, it was this moment that produced the truly heart-rending effect. No one had expected anything like this. With one voice we exclaimed 'Ah' and then there was a deadly silence. It was the moment in which God spoke and human beings listened attentively.

Previously, if the gradual fading and disappearance of nature had depressed us and made us desolate, and if we had imagined that process as a waning into a kind of death, we were now suddenly startled and jolted upright through the terrible power and violence of the movement which suddenly burst through the whole sky: the clouds on the horizon, which had earlier made us fearful, helped more than ever to produce the phenomenon. They now stood upright like giants, a terrible red hue ran from their vertex, and below they arched in a deep, cold, heavy blueness depressing the horizon. Banks of fog which had long welled up at the outermost edge of the Earth, and had merely been discoloured, now asserted themselves and shivered in a delicate, yet terrible glow which flooded them. Colours which no eye had ever seen coursed through the heavens.

The Moon was positioned in the centre of the Sun, no longer as a black disc, but in a semi-transparent state as if covered in a delicate steely gleam. The Sun had no edge but a wonderful, beautiful, shimmering ambience, bluish and reddish in colour, with broken beams, as if the Sun in its elevated position was pouring its wealth of light on the lunar sphere so that it sprayed in every direction — the most graceful effect of the light I have ever seen!

Extending far over the Marchfeld area, there lay a slanted, long pyramid of light with a horrible yellow hue, flaming in sulphurous colours and with an unnatural blue hem. It was the illuminated atmosphere beyond the shadow, and I have never seen a light that was so unearthly and so terrible, yet it provided the means by which we could see. If we had been put in a desolate mood by the previous monotony, we were now crushed by power and radiance and mass — our shapes were caught in the light like black, hollow spectres without any depth; the phantom of St Stephen's Cathedral was suspended in the air, the rest of the city was like a shadow, all the rattling had stopped, there was no longer any movement on the bridge; because each carriage and each rider had come to a halt and every eye was looking heavenward.

I will never, never forget those two minutes — they represented the powerlessness of a giant body, our Earth. How holy, how incomprehensible and how terrible is that entity which floods around us, which we soullessly enjoy and which causes our globe to tremble with such frisson when it withdraws its light for a short period. The air became cold, noticeably cold, dew fell so that clothes and instruments became damp, animals were terrified. The most awful thunderstorm was but superficial noise against such deathly silent majesty — the words of Lord Byron's poem came into my mind: the darkness where people set houses on fire, set forests on fire only to see light — but there was also such grandeur, I might almost say closeness to God, in these two minutes that the heart felt that he must be close by.

Byron was not enough — suddenly the words from the Holy Bible came into my mind, the words at the death of Christ: 'The Sun was darkened; the Earth did quake, and the rocks rent; the graves were opened; and the veil of the Temple was rent in twain from the top to the bottom.'

The effect on people's hearts was also evident. After the first fear had subsided, there were unarticulated sounds of wonder and amazement: some people lifted their hands, some wrung them quietly in front of them, others took and squeezed each other by the hands — one woman began to cry violently, another in the house next to us fainted, and one man, a serious and robust person, told me later that tears had been running down his face.

I always considered the old descriptions of solar eclipses to be an exaggeration just as this one might be considered as exaggerated in the future; but all of them, just like this one, do not do reality justice. They can only paint a picture of what is on show, and do so badly, and they are even worse at describing the associated feelings. What they are totally incapable of describing is the namelessly tragic music of colour and light that spread across the sky — a *Requiem*, a *Dies irae* which breaks our heart so that God sees it and his precious dead, so that he hears the call: 'Lord, how great and magnificent are your works, we are but dust before you because you can destroy us merely by puffing away a light particle and can transform our world, the comely and familiar place where we live, into an alien space in which larvae stare!'

But just as everything in creation has its allotted time, so too this phenomenon. Fortunately it only lasts for a very short time, a lifting of the cloak about his figure allowing us a fleeting access, only to cover himself up again so that everything returns to its previous state.

Just as people were beginning to express their feelings in words, that is, when they began to abate, when people began to exclaim: 'How magnificent, how terrible' — just at this moment it stopped: at once the alien world disappeared and the familiar one returned. A single droplet of light welled up at the upper edge like white hot metal and we had our world back. It thrust itself out as if the Sun itself were glad that it had overcome, a beam shot through space, a second appeared — but before people had time to call out 'Oh!' the larval world had disappeared at the flash of the first atom and our world had returned: and the lead-coloured horrific light, so frightening before its disappearance, now returned to refresh us, as friend and acquaintance, objects cast shadows again, the water sparkled, the trees were green once more, we looked each other in the eyes — ray followed victorious ray, and however small and tiny the first bright circle was, it appeared to us that we had been given an ocean of light. It is beyond expression, and anyone who has not experienced it will not believe the joyous, the victorious relief in our hearts: we shook hands, we said we would remember what we had seen for the rest of our days. We heard individual sounds as people called to one another from roofs and across the alleys, the driving and noise began again, even the animals felt it; horses neighed, the sparrows on the roofs began a clamour of joy as lurid and clownish as they usually are when very excited, and the swallows shot past and flashed up and down through the air.

The growing light no longer had any effect. Almost no one waited for the end, instruments were dismantled, we climbed down and in all the streets and pathways, groups and processions on their way home were engaged in the most animated and exalted discussions and exclamations. And even before the waves of admiration and veneration had subsided, before people could discuss with friends and acquaintances how the phenomenon had affected this person or that, here or there, the beautiful, graceful, warming, sparkling atmosphere returned, and the day's labour continued.

But for how long people's hearts continued to be agitated until they too could return to the day's labour — who can say? May God permit the impression to last; it was a magnificent one which little can challenge even though a person might live to a hundred years of age. I know that I have never been as affected, neither by music nor poetry, nor any other phenomenon of nature or art. Of course I have been familiar with nature since childhood and my heart is used to its language, and I love its language, perhaps more one-sidedly than is good; but I believe that there cannot be any heart in which this phenomenon has not left an indelible impression.

The Egyptian eclipse of August 30, 1905

M. WILHELM MEYER

Slowly the dark disc thrust its way further into the Sun. It was like inexorably approaching destiny. When the Sun's sickle looked no more than the three-day old Moon and there was still about fifteen minutes before the big moment, all those who had no reason to be there were asked to leave the area. There was just ourselves, our instruments and the waning Sun. A solemn silence descended; the Nile, too, flowed past us in solemn silence. A strange phenomenon appeared. The spots of light which shone through the palm leaves now took on a sickle shape and all the ground near our instruments displayed this strange pattern. About ten minutes before totality, a clear reduction in light intensity was noticeable on the landscape. I had set up my normal photographic equipment next to the telescope and now took a photograph of the landscape, leaving aperture and speed the same as for full sunlight.

Professor West, our commander general called: 'Five minutes to go, gentlemen.' We went to our posts. The light began to fade at a faster and faster rate. Only a few reflections of the small solar sickle still bubbled on the Nile almost like moonshine, and yet unlike it. It was a type of illumination that I had not seen before, and I sought in vain for comparisons to describe what I had had to describe hundreds of times previously, without really having seen it. One could say it was like an approaching thunderstorm though it was not the same yellowish light, but rather greyish blue. Also, the remaining sunshine was very weak, whereas an approaching thunderstorm tends to produce very sharp contrasts. It really was as if the whole of nature had been overcome by some kind of impotence, or rather, as if our sight was beginning to fade in these minutes, because nothing changed in the sky and on the Earth other than the light.

'One minute to go; are you ready, gentlemen?' Only the countdown rhythmically interrupted the eerie silence.

During the last ten seconds, the darkness increased with frightening speed. When the last rays of the Sun, spilling over the mountainous edge of the Moon, had died away, such a complete change in the scenery happened in the final seconds, that it came as a complete surprise, and affected me deeply within my being. It was as if nature had been fractured. If it had been dark previously, it now became wholly black in these first moments, like blackest night until our eyes became adjusted. Just as suddenly, as if the mysterious gleam from another world had begun to shine through during the previous second, the silver crown of the corona rays appeared; it was as if this light was emerging from the spot where the Sun had now completely disappeared and was being hurled at speed into dark outer space. As the place where the Sun had previously been now possessed the same darkness and colouring as the rest of the sky, and even appeared a little darker through the contrast with the corona, a confusing impression arose as if the Sun really had come from another world, dissolved into nothing, and had left behind this spectral glow around the emptiness it created in its wake, and as if the whole of nature possessed no more than just such a shadow existence. There was a cheerless orange-yellow glow on the horizon, caused by sections of the atmosphere no longer affected by the umbra of the Moon. This yellow light was also reflected in our faces so people looked like sallow shadows. If I recall, an artist created the picture of the landscape at the great moment described here.

The pulses of physical nature faltered, it seemed to be coming to a halt. We can be sure that everyone, even the most mindless person, stopped in their tracks as the lunar shadow raced over them. A characteristic incident in this respect occurred when the driver of a train travelling a few kilometres from Aswan station, confused by the impression that darkness was falling, stopped the train as if he was about to hit an obstacle. Our servant, who had remained at the inner sanctum of our station and who until then had been standing with arms crossed, suddenly ducked as if the thought something was about to fall on him out of the sky; he was about to run away, but in the face of the surrounding scientific calm he regained his composure and suffered the event to the end with arms crossed.

I only had a few minutes to let the impression of this great spectacle to sink in — it aroused a number of the strangest feelings. ... My eyes had adjusted and my senses, deeply affected at the beginning, had calmed down. I was clearly able to see two red flames, protuberances gleaming over the edge of the Moon, with my bare eyes, and follow some bundles of rays from the corona in their peculiar form, until they were about one-and-a-half times the diameter of the Sun in the sky. Some stars shone in the sky, Venus in particular.

Before we knew it, however, much faster than two minutes normally last, the first ray of Sun flashed up from the edge of the Moon and the corona withdrew back into itself; normal daylight seemed to return much faster than it had disappeared.

All of us took a deep breath. ... What secrets had the heavens revealed to us?

Lunar eclipses 2005–2017

Date	Time (GMT)	Type	Page
2005 Oct 17	12.04	Partial	
2006 Sep 7	18.52	Partial	
2007 March 3	23.22	Total	149
2007 Aug 28	10.37	Total	152
2008 Feb 21	03.26	Total	154
2008 Aug 16	21.11	Partial	
2009 Dec 31	19.24	Partial	
2010 June 26	11.40	Partial	
2010 Dec 21	08.17	Total	156
2011 June 15	20.13	Total	158
2011 Dec 10	14.32	Total	160
2012 June 4	11.04	Partial	
2013 April 25	20.10	Partial	
2014 April 15	07.46	Total	162
2014 Oct 8	10.55	Total	164
2015 April 4	12.02	Total	166
2015 Sep 28	02.48	Total	168
2017 Aug 7	18.22	Partial	

Lunar eclipse of September 1996. The light rim at the top shows the end of totality approaching. The reddish colour is caused by the high dust content of the Earth's atmosphere.   *(Photo: Paul Mortfield)*

The lunar eclipses

Character and nature of a lunar eclipse

During a lunar eclipse, it is the Earth itself which casts its shadow on its satellite. Since the Moon is only 27% of the size of the Earth, the whole of the Moon is fully immersed in darkness. But this darkness is not absolute because, at the edge of the Earth, the sunlight is refracted by the thin atmosphere, and scattered into the 14 000 000 km long shadow cone of the Earth as a reddish to brownish grey glow. Accordingly, the Moon does not completely lose its shine when it dips into the shadow but continues to shimmer in a copper red, sallow light. No other celestial phenomenon has such a mysterious and mystical effect as the sombre, matt lunar disc during a lunar eclipse.

A lunar eclipse is also the opposite of a solar eclipse in terms of its progression over time: in a solar eclipse events occur in sudden, dramatic changes, but the lunar eclipse takes place gradually. The scattering of the sunlight in the Earth's atmosphere means that there is no sharply drawn line of shadow on the Moon, but a diffused transition from the brightly lit part into darkness, covers increasingly large areas of the moonscape. This spectacle is peculiar and disconcerting because the Moon appears in a way which contrasts with its normal appearance in two respects: its cool and clear silvery light is turned into a matt red copper to brown shimmer and the sharp division of light and shadow which characterizes the Moon has been transformed into an indistinct transition. Those elements which precisely characterize the Moon and represent its particular features, are lost during the eclipse.

If we understand the outer appearance of a celestial body as an expression of its inner nature, we begin to see that much more happens during a lunar eclipse than simply a temporary change in colour of the Earth's satellite. It then becomes important, for instance, for a more profound understanding of this celestial spectacle, that the fact that it is always the Full Moon which is eclipsed, should not simply be discounted as pure spatial necessity. What is the Full Moon? Among the various lunar phases the Full Moon marks the position in which the Moon most strongly comes to expression. At the same time it is during Full Moon that most of the lunar terrestrial influences such as the tides or water balance of plants become evident. The small lunar sickle at dusk primarily addresses our feelings, but the Full Moon displays the side of the Moon which comes to expression in volition. And it is precisely this aspect of the Moon which is eclipsed.

We can sense that the Moon is 'cast into shadow' during this period not just visually, simply forfeiting its light, but that other aspects of its character are also 'eclipsed.'

It is a feature of the Moon that, in several respects, it represents a border between terrestrial space and the planetary system. For example, the Earth's magnetic field extends to the Moon. The idea also recurs in the cosmological concepts of antiquity about the hierarchical structure of the planetary system that the Moon, or rather the space it prescribes, the sublunar sphere, should be considered as a transition from terrestrial to planetary space. It is not just that formations of the moonscape are visible to the naked eye, the distance of the Earth's satellite at 380 000 km is still within 'human' measure. The amount we walk throughout our lives approaches such distances. It is no surprise, then, that in accordance with early Christian cosmological ideas, this space was considered to be the home of those supersensory beings, which in turn form the bridge and are mediators between human beings and the divine world: the angels.

Observation of a lunar eclipse offers the opportunity to see whether alongside the eclipse of the lunar light these delineating characteristics of the Moon are also weakened and whether as a result the Earth is without protection in a certain sense for a short period of time. Rudolf Steiner described the consequences of such a dropping of the lunar boundaries in dramatic terms:

'When there is an eclipse of the Moon modern consciousness says, here we have the Earth between the Sun and the Moon and that is why we see the Earth's shadow on the Moon. That is a physical explanation. But the ancient initiates knew that there is an underlying spiritual element; they knew that in being eclipsed, thoughts stream down through the darkness, that they therefore have a more intimate relationship to the human subconscious than to consciousness. And the ancient initiates often told their pupils in the form of an analogy — I translate into modern language — romantic people take a stroll by the light of the Full Moon; but those people who want to absorb evil thoughts from the cosmos, not good thoughts, take a walk during a lunar eclipse.'*

Progression of a lunar eclipse

Since the lunar trajectory is inclined in relationship to the ecliptic, the Full Moon mostly travels above or below the conical shadow of the Earth as described. Twice a year on average, the particular situation occurs that the Moon is in the plane of the ecliptic during Full Moon. It is positioned close to one of its two nodes. Then the Moon is immersed in the conical shadow zone

* Human Questions and Cosmic Answers, lecture of June 25, 1922.

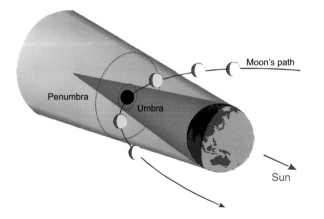

The Moon enters the penumbra before moving through the umbra, and exiting through the penumbra again.

extending 14 000 000 km beyond the Earth. First it enters the penumbra of the Earth. At this point, an observer on the Moon would see how the Earth covers a part of the solar disc. The Moon looses only a little light as it progresses through the penumbra so that only experienced observers see a reduction in brightness on the moon's surface. After about an hour the Moon enters the umbra.

The diffuse border of the shadow can now be seen on the Moon. It takes on a pale brown to reddish copper colour. After a maximum of 2.5 hours of total lunar eclipse the Moon detaches itself from the Earth's shadow and gradually regains its original luminosity.

2007 March 3–4, total lunar

Whereas only partial lunar eclipses occur in 2005 and 2006, this lunar eclipse in March once again introduces a series of total lunar eclipses. This one, occurring shortly before the spring equinox, is a European and African eclipse because that is where it can be observed in full. Naturally it takes place in the sign of the zodiac opposite the Sun, Leo. Not far from the Full Moon, Saturn is positioned in the same constellation.

Both luminaries, the Moon and Saturn, have an interesting correlation in the period of their trajectory: the Moon takes 29.5 days from one Full Moon to the next. Saturn takes 29.5 years for its journey through the zodiac. Spiritually oriented astronomy therefore rightly ascribes characteristics to Saturn which are related to the Moon. Just as the Moon forms a border between the terrestrial and planetary sphere, so Saturn forms a threshold between the classic planetary system and the world of the stars. Beyond that, the relationship between days and years, as reflected here in the period taken for one full revolution, is well known from the situation on Earth. The atmospheric features within the course of the day can be found again in the corresponding periods of the year. Morning in the day is equivalent to spring in the year. Correspondingly, midday reflects summer, afternoon the autumn and night winter. Thus Christmas is a winter festival, a festival of night, and Easter a festival of the morning. An active experience of the seasons of the year therefore also includes the enjoyment of the seasons as enhanced by the appropriate time of day. Thus spring can be most purely understood at around sunrise, summer in the midday period, autumn during the late afternoon and winter in the night. These temporal correlations can be seen spatially in Saturn and Moon.

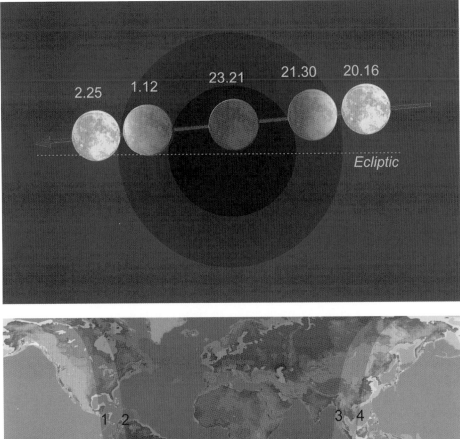

Overview of the lunar eclipse:
1 End of total eclipse at moonrise (sunset)
2 Beginning of total eclipse at moonrise (sunset)
3 End of total eclipse at moonset (sunrise)
4 Beginning of total eclipse at moonset (sunrise)

2007 March 3–4, Sat/Sunday	Date	Time (GMT)
Entry of Moon into penumbra:	March 3	20.16
Entry of Moon into umbra:	March 3	21.30
Start of total eclipse:	March 3	22.44
Maximum of eclipse:	March 3	23.21
End of total eclipse:	March 3	23.58
Moon leaves umbra:	March 4	01.12
Moon leaves penumbra:	March 4	02.25

2007 August 28, total lunar

The duration of total darkness during this lunar eclipse is relatively long at 1 hour and 31 minutes. This is due to the fact that the Full Moon and the Moon crossing the ecliptic are relatively close together. At 06.28, that is only four hours before the middle of the eclipse, the Moon reaches its node. This means that it travels through the Earth's shadow almost at its widest place. The eclipse is visible from western North and South America (night of August 27/28), Japan and Australasia (night of August 28/29).

2007 Aug 28, Tuesday	Date	Time (GMT)
Entry of Moon into penumbra:	Aug 28	07.52
Entry of Moon into umbra:	Aug 28	08.51
Start of total eclipse:	Aug 28	09.52
Maximum of eclipse:	Aug 28	10.37
End of total eclipse:	Aug 28	11.23
Moon leaves umbra:	Aug 28	12.24
Moon leaves penumbra:	Aug 28	13.22

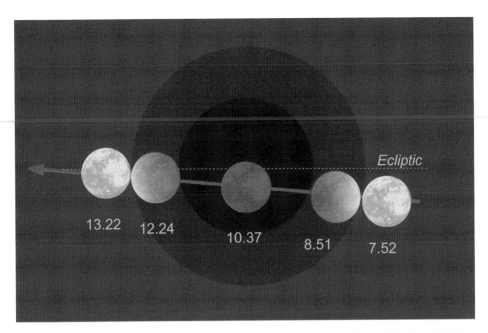

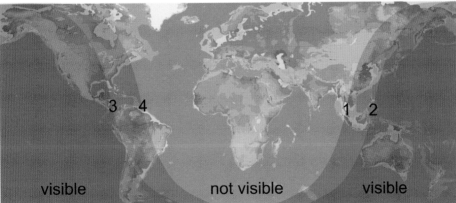

Overview of the lunar eclipse:

1 End of total eclipse at moonrise (sunset)

2 Beginning of total eclipse at moonrise (sunset)

3 End of total eclipse at moonset (sunrise)

4 Beginning of total eclipse at moonset (sunrise)

2008 February 21, total lunar

This lunar eclipse can be seen from North and South America, Europe and most of Africa. In Leo, the Moon travels past Regulus, the main star of this strong and easily recognizable constellation, just after midnight 1° below the star. One hour later it moves into the Earth's shadow. From a European perspective, it leaves the umbra before dawn. The direct presence of Regulus on the right side and Saturn on the left side gives the spectacle particular relevance. At the same time the powerful language of forms of the constellation of Leo makes it especially dramatic. This lunar eclipse could with some justification be described as one of the most impressive in this decade.

2008 Feb 21, Thursday	Date	Time (GMT)
Entry of Moon into penumbra:	Feb 21	00.35
Entry of Moon into umbra:	Feb 21	01.43
Start of total eclipse:	Feb 21	03.00
Maximum of eclipse:	Feb 21	03.26
End of total eclipse:	Feb 21	03.51
Moon leaves umbra:	Feb 21	05.09
Moon leaves penumbra:	Feb 21	06.17

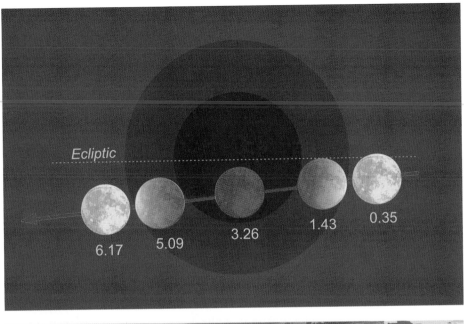

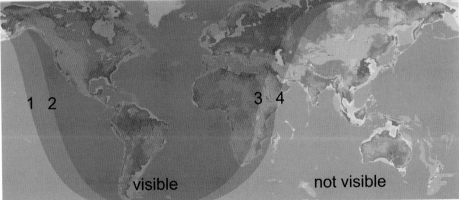

Overview of the lunar eclipse:

1 End of total eclipse at moonrise (sunset)

2 Beginning of total eclipse at moonrise (sunset)

3 End of total eclipse at moonset (sunrise)

4 Beginning of total eclipse at moonset (sunrise)

2010 December 21, total lunar

The remarkable thing about this eclipse is that it takes place just one day before the winter solstice. In the northern hemisphere, the Sun is in its lowest position of the year, while in contrast the Moon is in its highest position beetween Taurus and Gemini. No Full Moon climbs so high in the night sky, no Full Moon dominates celestial events to such an extent as the Full Moon of the winter solstice. For the observer, this creates the impression as if the Moon were standing almost in the zenith. (In fact it would only stand there when observed from the Tropic of Cancer, such as in Mexico.) In this lunar eclipse it is the high Full Moon of the winter solstice which is darkened, and it is this Full Moon which is prevented from developing its activity. Only actual observation will confirm this, but we may assume that this lunar eclipse will be of particular intensity. The last time that the Moon was eclipsed at the time of the winter solstice was in 1638.

The eclipse can be observed from North and South America in the evening of Monday, December 20.

2010 Dec 21, Tuesday	Date	Time (GMT)
Entry of Moon into penumbra:	Dec 21	05.28
Entry of Moon into umbra:	Dec 21	06.32
Start of total eclipse:	Dec 21	07.40
Maximum of eclipse:	Dec 21	08.17
End of total eclipse:	Dec 21	08.53
Moon leaves umbra:	Dec 21	10.02
Moon leaves penumbra:	Dec 21	11.06

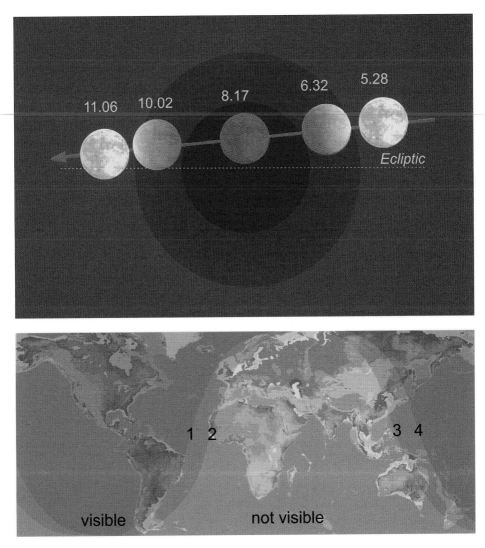

Overview of the lunar eclipse:

1 End of total eclipse at moonrise (sunset)

2 Beginning of total eclipse at moonrise (sunset)

3 End of total eclipse at moonset (sunrise)

4 Beginning of total eclipse at moonset (sunrise)

2011 June 15, total lunar

Africa, the Middle East, Asia and Australia are the areas of the world from which this lunar eclipse can be observed. With its totality lasting 1 hour 41 minutes, it is very long. The reason for this is that Full Moon and the passage through the node are less than two hours apart, which means that the Moon travels through the centre of the Earth's shadow.

The position of the Moon in the zodiac is worthy of note. It is positioned above the tail of Scorpio. If we observe this region of the stars carefully, we can feel the extraordinary tension existing in this region of the zodiac: whereas Scorpio possesses a dynamic S-shaped form and asserts itself with a bright, almost aggressive flourish, we see above it (below in southern hemisphere) the large, round, weakly glimmering shape of the Serpent Bearer. What reveals itself morphologically in the starry sky, is reflected in the myths of antiquity: the scorpion, representative of death and consciousness, stands in confrontation with the Serpent Bearer, Asclepius, the demigod of healing. Death and life, consciousness and dreams are the combinations of ideas which clash in this region of the zodiac and which are now particularly emphasized through the obscured Moon.

2011 June 15, Wednesday	Date	Time (GMT)
Entry of Moon into penumbra:	June 15	17.23
Entry of Moon into umbra:	June 15	18.23
Start of total eclipse:	June 15	19.23
Maximum of eclipse:	June 15	20.13
End of total eclipse:	June 15	21.04
Moon leaves umbra:	June 15	22.04
Moon leaves penumbra:	June 15	23.04

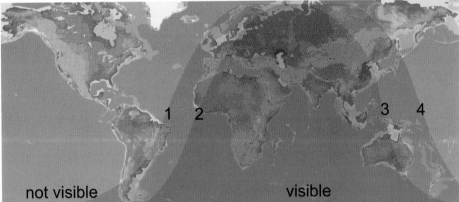

Overview of the lunar eclipse:
1 *End of total eclipse at moonrise (sunset)*
2 *Beginning of total eclipse at moonrise (sunset)*
3 *End of total eclipse at moonset (sunrise)*
4 *Beginning of total eclipse at moonset (sunrise)*

2011 December 10, total lunar

The small V-shaped pattern of stars at the centre of Taurus is the most distinctive shape of this winter constellation. Two extensive rows of stars reach out towards the left from this compact shape and lend the image particular dynamism and power. The Moon is positioned between these two ray-like extensions of the constellation — normally described as the horns of the bull — as it enters the shadow of the Earth. Further afield we have Gemini, Orion and Boötes, the Waggoner, so that the obscured Moon is surrounded by starlight. Mars and Jupiter, respectively positioned one and two constellations of the zodiac distant, reinforce the impression of a particularly 'bright' lunar eclipse.

As it already passed through the descending node 7.5 hours before the Full Moon position, it travels through the Earth's shadow a little below the ecliptic.

The eclipse is visible in Asia, Australasia and north-west North America.

2011 Dec 10, Saturday	Date	Time (GMT)
Entry of Moon into penumbra:	Dec 10	11.32
Entry of Moon into umbra:	Dec 10	12.46
Start of total eclipse:	Dec 10	14.06
Maximum of eclipse:	Dec 10	14.32
End of total eclipse:	Dec 10	14.59
Moon leaves umbra:	Dec 10	16.17
Moon leaves penumbra:	Dec 10	17.32

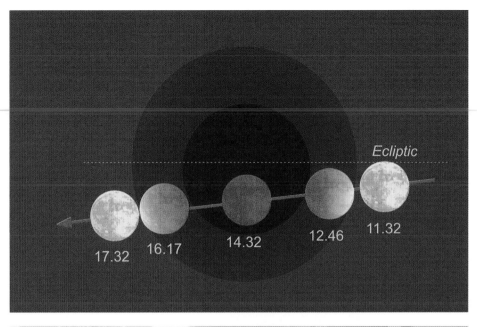

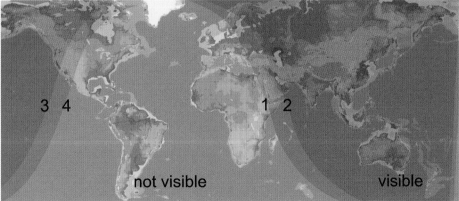

Overview of the lunar eclipse:
1 End of total eclipse at moonrise (sunset)
2 Beginning of total eclipse at moonrise (sunset)
3 End of total eclipse at moonset (sunrise)
4 Beginning of total eclipse at moonset (sunrise)

2014 April 15, total lunar

Only partial lunar eclipses occur in 2012 and 2013 as well as the hardly noticeable penumbra eclipses, but two total lunar eclipses take place in 2014. In mid-April, the first one in the year can be seen from South and North America in the night of April 14/15.

Spica, the main star in Virgo, is positioned directly south of the Moon which takes on a red to greyish brown colour. The blue radiating shine of Spica will form an attractive contrast to the sallow light of the dark Moon. In addition, Saturn in Libra and Mars in Virgo form a planetary frame around the lunar eclipse.

2014 April 15, Tuesday	Date	Time (GMT)
Entry of Moon into penumbra:	April 15	04.53
Entry of Moon into umbra:	April 15	06.00
Start of total eclipse:	April 15	07.07
Maximum of eclipse:	April 15	07.46
End of total eclipse:	April 15	08.26
Moon leaves umbra:	April 15	09.34
Moon leaves penumbra:	April 15	10.40

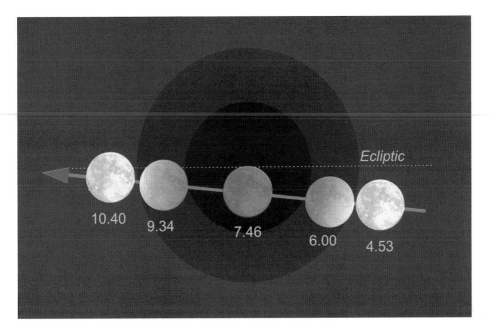

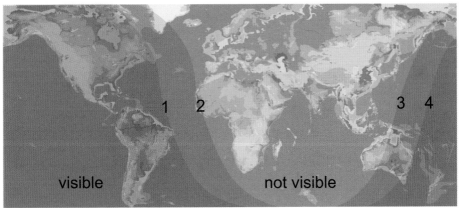

Overview of the lunar eclipse:

1 *End of total eclipse at moonrise (sunset)*

2 *Beginning of total eclipse at moonrise (sunset)*

3 *End of total eclipse at moonset (sunrise)*

4 *Beginning of total eclipse at moonset (sunrise)*

2014 October 8, total lunar

Full Moon in October occurs seven hours before the descending lunar node so that the Moon is still above the ecliptic at the time of the eclipse, correspondingly travelling through the Earth's shadow above the centre. The eclipse can be seen particularly from the Pacific. North and Central America as well as the north-western part of South America provide a 'firm' place of observation on the evening of October 7; New Zealand, Australia and north-west Asia will see the eclipse at dawn of October 8.

In October, the Full Moon is positioned in Pisces. The faint stars of this constellation are obscured by the moonlight. It will probably be possible to observe well how the stars of this large constellation become brighter during the eclipse. The only time that the Full Moon and its backdrop in the constellation of Pisces can be observed at the same time is during the lunar eclipse.

2014 Oct 8, Wednesday	Date	Time (GMT)
Entry of Moon into penumbra:	Oct 8	08.14
Entry of Moon into umbra:	Oct 8	09.14
Start of total eclipse:	Oct 8	10.25
Maximum of eclipse:	Oct 8	10.55
End of total eclipse:	Oct 8	11.24
Moon leaves umbra:	Oct 8	12.35
Moon leaves penumbra:	Oct 8	13.35

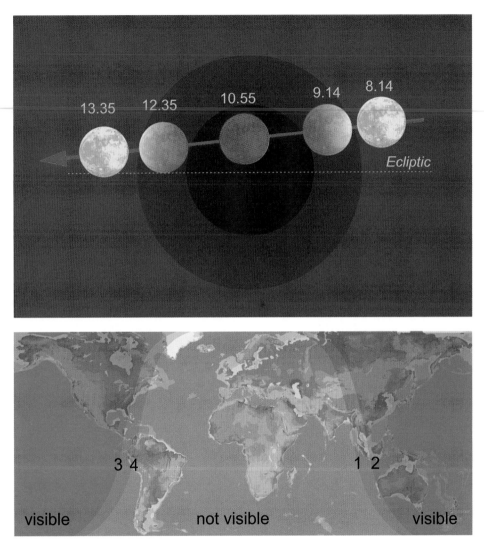

Overview of the lunar eclipse:

1 End of total eclipse at moonrise (sunset)
2 Beginning of total eclipse at moonrise (sunset)
3 End of total eclipse at moonset (sunrise)
4 Beginning of total eclipse at moonset (sunrise)

2015 April 4, total lunar

The eclipse of April 4, is rightly described by Philip Harrington as the brightest because the Moon travels past the extreme northern edge of the Earth's shadow. Thus, the actual eclipse only lasts about 12 minutes. There has not been such a short period of totality for more than a hundred years. Scattered light from the Earth will undoubtedly illuminate the other side of the Moon. Spica, the main star in Virgo, will also contribute to the impression of brightness. The whitish blue star is positioned close to the obscured Moon forming a strong contrast to its darkly ochre colouring.

The eclipse can only be seen in full around the Pacific. From the North American west coast the partially obscured Moon can be watched as it sets on April 4, while in Asia and Australia, it will only be visible at dawn on Sunday, April 5.

2015 April 4, Saturday	Date	Time (GMT)
Entry of Moon into penumbra:	April 4	09.00
Entry of Moon into umbra:	April 4	10.16
Start of total eclipse:	April 4	11.55
Maximum of eclipse:	April 4	12.01
End of total eclipse:	April 4	12.07
Moon leaves umbra:	April 4	13.44
Moon leaves penumbra:	April 4	15.01

Overview of the lunar eclipse:

1 End of total eclipse at moonrise (sunset)

2 Beginning of total eclipse at moonrise (sunset)

3 End of total eclipse at moonset (sunrise)

4 Beginning of total eclipse at moonset (sunrise)

2015 September 28, total lunar

This eclipse, which can be observed from Europe, the east coast of North America, South America and Africa, occurs shortly after the autumnal equinox. As in the previous year, the Moon is positioned in Pisces in the zodiac. While the distinctive quadrilateral constellation of Pegasus shines above the Moon, the stars in Pisces cannot compete with the light of the Full Moon. Only when the Moon enters the Earth's shadow and increasingly loses its luminosity, do the weakly shining stars of Pisces, forming two attractive bands of light around Pegasus, appear.

It is worthy of note that perigee, the point when the Moon is closest to the Earth during its trajectory, occurs only one hour before Full Moon. As a result the diameter of the Full Moon appears to be bigger by $1/7$ than when the Moon is at apogee.

2014 Sep 28, Monday	Date	Time (GMT)
Entry of Moon into penumbra:	Sep 28	00.11
Entry of Moon into umbra:	Sep 28	01.08
Start of total eclipse:	Sep 28	02.12
Maximum of eclipse:	Sep 28	02.48
End of total eclipse:	Sep 28	03.25
Moon leaves umbra:	Sep 28	04.29
Moon leaves penumbra:	Sep 28	05.25

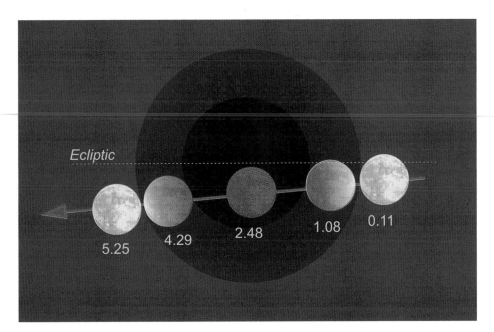

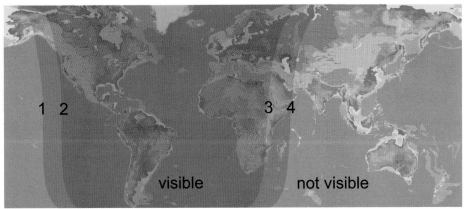

Overview of the lunar eclipse:

1 End of total eclipse at moonrise (sunset)

2 Beginning of total eclipse at moonrise (sunset)

3 End of total eclipse at moonset (sunrise)

4 Beginning of total eclipse at moonset (sunrise)

Special astronomical events

2005 Sep 2. Jupiter, Venus and Spica

In the early dusk of the first days of September, a picturesque grouping made up of the two brightest planets, Venus and Jupiter, together with Spica, the bluish white main star of Virgo, is visible. From temperate latitudes the constellation can only be observed a few degrees above the landscape so that an unrestricted view of the western horizon is necessary. Encounters between Jupiter and Venus belong to the most impressive conjunctions because of the brightness of the two planets. Although Venus with its lavish light is clearly brighter than Jupiter, the latter planetary giant nevertheless appears to possess greater radiance. The presence of the bright star, Spica, enhances the planetary encounter. The star appears to lend something of its eternity to the brief encounter of the planets.

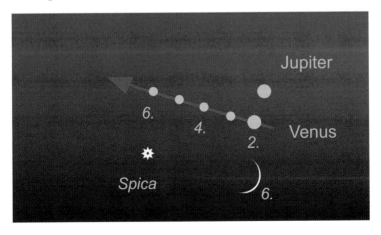

Evening view westwards, 2005 September 2–6

2005/6.
Saturn's loop in the Manger: Death and birth

Saturn takes almost 30 years for a full revolution and so stays an average of two-and-a-half years in each constellation of the zodiac. In 2006, the distant planet is in the constellation of Cancer making its annual loop directly in the centre of the constellation. There we can also see the open cluster of stars called Praesepe (the manger) or also M44. The name Manger comes from the idea in antiquity that the human soul in its prenatal journey towards Earth enters the zodiac in Cancer. The shape of this constellation corresponds to the motif of 'coming down to Earth': on three sides weaker stars form a concentric movement towards a middle field which in turn is marked by three stars. This cluster of stars is positioned very close to the centre of the three rays heading towards the middle, and to the bare eye it appears like a faded spot of mist, which in binoculars swells into a multitude of stars. In 2006 it is Saturn which lends emphasis to this 'constellation of birth.' This creates an interesting contrast because Saturn represents the other side of life: inwardness, maturity and also death. Saturn first travels past this cluster of stars in September 2005, then in February of the following year and finally for the third time in June.

The intimate and recurring view of this image of the planet and the constellation of the zodiac, which lends its character to the whole of 2006, can enhance our understanding of the common characteristics of birth and death, perhaps in the sense expressed by Goethe when he said: 'Death is the invention of life to gain more life.'

In addition there is a noteworthy conjunction between Mars and Delta Cancerii on June 17. All three luminaries are positioned in a row and can be seen at dusk. The contrasting colours between Mars and Saturn add to the attraction.

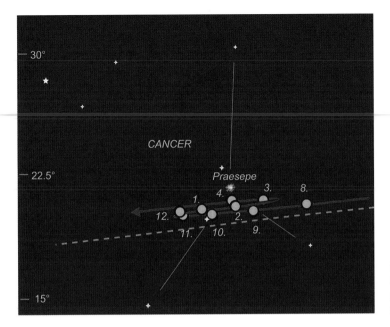

30°

CANCER

22.5°

Praesepe

1.
4.
3.
8.
12.
2.
11.
10.
9.

15°

Saturn's loop in Cancer.

2005 August to 2006 September

2008 July 6.
Royal assembly of Mars, Saturn, Moon in Leo

The encounter of two planets occurs frequently, however, an assembly of three wandering stars is a rare event. This applies all the more if such a conjunction takes place near a bright star of the zodiac. On July 6, it is worth looking up into the western evening sky shortly after sunset. Above the small waxing lunar sickle Saturn, Mars and Regulus, the main star in Leo, can be seen like a three-pointed crown.

There are interesting relationships and contrasts among these four lights: Leo, with Regulus as its main star, is the constellation of the zodiac which possesses the dynamism, combative and expressive power which Mars occupies among the planets. Anyone who studies the language of forms of the constellations of the zodiac in greater depth will therefore be able to see Leo as a suitable 'house' for the red planet. Just as Regulus and Mars are related, there is a close connection between the Moon and Saturn. This becomes evident, for example, in the extraordinary correlation of their rhythms: the Moon requires 29.5 days for its full synodic course and Saturn requires 29.5 years to complete its revolution. The outer appearance confirms the inner connection between the two. In the myths of antiquity, but also in their various forms of appearance, both represent inwardness, calmness and inner maturity — the Moon in relation to the Earth, Saturn majestically related to the stars. To this extent the contrasts of dynamism and inwardness, battle and inner peace encounter one another in twofold respect on this July.

LEO

Saturn

Mars

Regulus

7.

6.

West, 22.30

*Evening view
(22.30) westwards
from temperate
northern latitudes.*

2008 July 6–7

2009. The duet of Venus and Mars

If we look towards the east at dawn on April 21, 2009 it is easy to see bright Venus. It is more difficult to find Mars, almost ten Moon diameters lower. The conjunctions of the other planets always have the character of overtaking or catching up, but these two planetary neighbours, Mars and Venus, can travel together in tandem for weeks at a time. This now happens impressively in May and June: Venus has concluded its forward movement and now stands almost motionless against the stellar background for a number of days. As though waiting for Mars, it starts its motion again as the red planet approaches. It follows Mars through

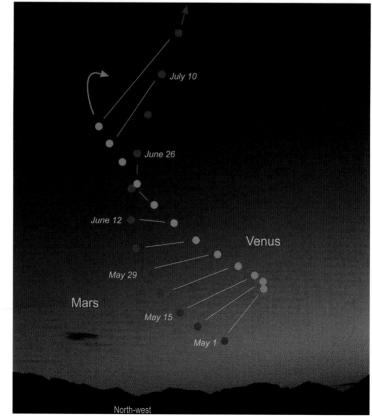

Overview of Mars and Venus

2009

May 1 – July10

SPECIAL ASTRONOMICAL EVENTS

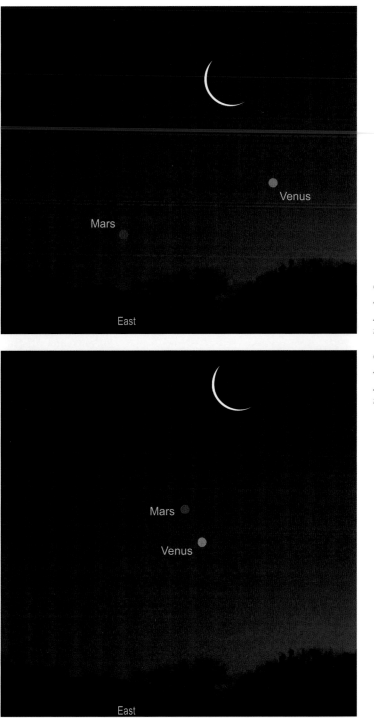

Conjunction of Mars and Venus

2009 May 21 morning

Conjunction of Mars and Venus

2009 June 19 morning

177

*Conjunction of
Mars and Venus
2009 July 19
morning*

Pisces at a constant distance of about 6°. On June 20/21, at the time of the summer solstice, it has caught up, and as if to confirm it the picturesque lunar sickle joins the two planets during these days. Now the process is reversed and Mars follows Venus, rushing ahead, but must accept an increasing distance throughout August. Rarely is it possible to observe a planetary encounter as such a long drawn-out process of waiting, approach, and pursuit as happens in the first half of 2009.

SPECIAL ASTRONOMICAL EVENTS

2010 August 14.
Venus, Moon, Mercury Mars and Saturn

This last summer of the first decade of the new millennium will be remembered for a special evening constellation. If we look towards the western evening sky on July 9, we will see an ascending series of three planets: Mars, Venus and Saturn. Mercury, which is also part of this line, cannot assert itself in the light of dusk. In addition, the Moon can be seen below these three planets on July 15. While Mars travels towards Saturn in the following days and weeks, Venus catches up with the red planet at the same time so that the three planets move closer and closer together. At the start of August, a few days before the Moon once again visits the planetary assembly, the planets form an attractive triangle. With the end of summer, all three disappear in the radiance of the Sun, and we are left with the rich impression of the complexity and subtlety of a conjunction of three planets in contrast to the conjunction of two.

Venus, Mars, Saturn and Moon in the south-west from northern temperate latitudes.

2010 August 13 evening

2012 June 5/6. The transit of Venus

On June 5/6, 2012, the rare event of a renewed transit of Venus will occur, following the one in 2004. During these days, the sun-oriented neighbour of the Earth will move directly across the Sun and be visible (with protected eyes) as a black dot moving across its disk. The beginning of the transit is visible from North America in the afternoon of June 5. Thus on the American west coast Venus will be visible as a circular shadow on the disc of the Sun at about 22.06 GMT (3.06 pm PDST). The difference in perspective means that the transit will begin four minutes later when viewed from Japan, China or Australia where it is already June 6, due to the date line.

In Europe the start of the transit will not be visible as it is shortly after midnight (CEST) when Venus begins to move across the Sun. Venus will leave the Sun again when the latter is posi-

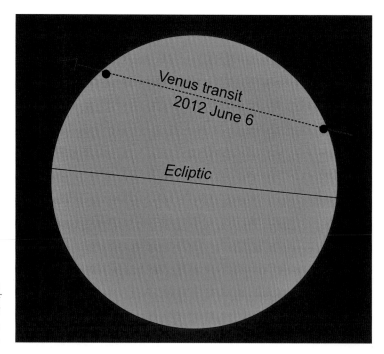

Transit of Venus across the disc of the Sun. Projected from a telescope onto white card.

SPECIAL ASTRONOMICAL EVENTS

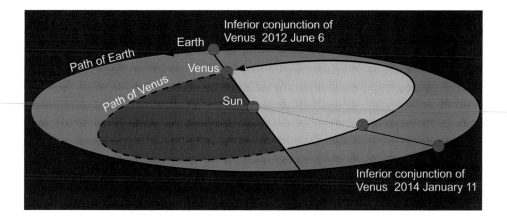

tioned about 10° above the horizon so that at least the end of the transit can be seen from Europe.

Inferior conjunctions of Venus

Venus is positioned in its inferior conjunction during this event. With a distance of 43 000 000 km it is at this point at the least distance possible to the Earth. During this position its apparent size has grown to one angular minute or $^1/_{30}$ of the diameter of the Sun. In this respect observation of the transit of Venus offers the opportunity of seeing a planet of the solar system at extraordinary 'closeness.'

A transit of Venus takes place when Venus is positioned both between the Earth and the Sun, that is in its inferior conjunction, while also being near its node. The nodes are the two locations in a planetary trajectory at which the planet crosses the plane of the Earth's trajectory, the ecliptic. Normally Venus is positioned below or above the Sun during a lower conjunction, like the New Moon, so that there is no transit.

On average, transits of Venus only take place twice in each century. The reason for this is that the five locations in the zodiac at which Venus is positioned in its lower conjunction only move very slowly. Venus takes eight years to etch the complete pentagram of the five lower conjunctions into the heavens. After this period it is once again positioned in lower conjunction with the slight difference of 2.3 days. The pentagram thus turns slowly in a clockwise direction. It takes 243 years for the pentagram to

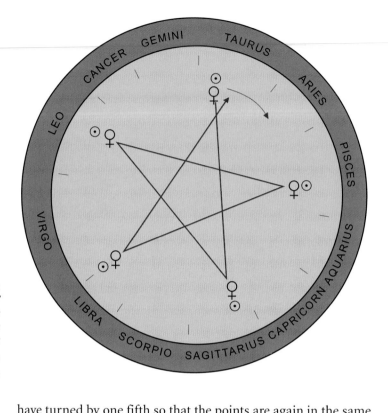

*The pentagram
of the Venus'
conjunctions slowly
rotate round the
zodiac*

☉ *Sun*

♀ *Venus*

have turned by one fifth so that the points are again in the same direction.

It is an element in the mysterious correlations of movement in the planetary system that Venus requires the same number of days for its rotation as is made up in years by the travel of the tips of the pentagram. This large-scale cycle of Venus' conjunctions means that the current conjunction locations correspond to those from 1760. There was indeed a transit of Venus in 1761. This is of historical importance as it was the first predicted transit of Venus, foretold by the astronomer Johannes Kepler (1571–1630).

Venus will not transit the Sun in central position but will be seen in the outer region since Venus descends across the ecliptic as early as June 7. The minor tilt of Venus's trajectory of 3.4° in relation to the ecliptic makes it possible for the time that Venus crosses the node and its precise position between Earth and Sun, the so-called lower conjunction, to lie 1.5 days apart.

Transits of Venus occur more frequently than just every 243 years. There are two reasons for this: on the one hand, transits occur both on the descending as well as the ascending node so that a rhythm results which is twice as frequent.

After eight years the location of the original conjunction is only missed by 2.3°, so there are an additional two transits of Venus in this eight year rhythm. In the first transit, passage through the node lies before the conjunction with the Sun, in the second transit after the conjunction. Four transits of Venus occur in the 243-year rhythm described, and they do so at the following intervals:

121.5 years — 8 years — 105.5 years — 8 years

Shifts over long periods of time, such as for instance the full revolution of the node in about 63 000 years, means that the transit rhythm is not constant. This rhythm of Venus only applies from the middle of the second millennium and will lose its validity again in the fourth millennium. The dates of the transits (June 7, December 12) will also gradually move back.

But the fourfold Venus cycle described above will apply for the next centuries. We are at the beginning of such a 243-year cycle because the last transit took place 122 years ago in December 1882, and Venus will not move across the Sun again until 2117.

The following times apply to the Venus transit on June, 6, which can diverge by up to seven minutes at various locations around the world, as previously indicated. As the entry and exit from the transit are particularly worth seeing, a number of cities have been listed below with precise times.

In Central Europe the Sun only rises at 5.30 in the summer, so only the final phase of the transit is visible there. This does, however, provide the opportunity to observe the passage of Venus without a Sun filter: when the Sun is situated behind sufficient haze on the horizon directly as it rises, its light has been reduced

Venus transit, 2012 June 5/6		(degrees after time show height of Sun)			
All times in GMT	touches edge of Sun	fully in Sun	central	exits Sun	totally out of Sun
Adelaide, Australia	22.16.01 4°	22.34.08 7°	01.30.59 30°	04.27.00 27°	04.44.58 26°
Amsterdam, Netherlands				04.37.12 9°	04.54.51 11°
Atlanta, Georgia, USA	22.04.17 31°	22.21.56 27°			
Athens, Greece				04.37.50 16°	04.55.26 20°
Auckland, New Zealand	22.15.25 24°	22.33.27 25°	01.29.1 28°	04.25.08 7°	04.43.17 4°
Barcelona, Spain				04.37.45 2°	04.55.26 5°
Beijing, China	22.09.53 14°	22.27.40 17°	01.30.32 52°	04.31.51 7°	04.49.20 71°
Berlin, Germany				04.37.13 14°	04.54.50 16°
Bern, Switzerland				04.37.32 8°	04.55.11 11°
Birmingham, UK				04.37.07 5°	04.54.47 8°
Brisbane, Australia	22.15.43 19°	22.33.45 22°	01.30.02 40°	04.26.00 2°	04.43.59 24°
Brussels, Belgium				04.37.17 8°	04.54.56 10°
Caracas, Venezuela	22.04.37 9°	22.22.26 5°			
Chicago, Illinois, USA	22.04.12 34°	22.21.49 31°			
Denver, Colorado, USA	22.05.11 47°	22.22.47 44°	01.25.36 9°		
Dublin, Ireland				04.36.59 4°	04.54.40 6°
Glasgow, UK				04.36.51 6°	04.54.32 8°
Hamburg, Germany				04.37.09 12°	04.54.46 15°
Havana, Cuba	22.04.38 27°	22.22.20 23°			
Helsinki, Finland			01.29.58 1°	04.36.36 22°	04.54.24 24°
Honolulu, Hawaii, USA	22.09.59 5°	22.27.38 89°	01.26.12 49°	04.26.30 9°	04.44.29 5°
Ho Chi Minh City, Vietnam			01.32.06 41°	04.31.28 77°	04.49.01 78°
Jerusalem, Israel				04.37.40 24°	04.55.14 28°
Kansas City, Kansas, USA	22.04.37 39°	22.22.15 36°	01.25.44 2°		
Kingston, Jamaica	22.04.37 21°	22.22.21 17°			
Lima, Peru	22.06.54 9°	22.24.49 5°			
London, UK				04.37.12 6°	04.54.52 8°
Los Angeles, California, USA	22.06.17 58°	22.23.53 55°	01.25.24 18°		
Lusaka, Zambia				04.36.44 1°	04.54.33 5°
Lyon, France				04.37.34 6°	04.55.14 9°
Madrid, Spain					04.55.25 1°
Manchester, UK				04.37.02 6°	04.54.43 8°
Marseille, France				04.37.42 5°	04.55.22 8°
Melbourne, Australia	22.16.03 7°	22.34.10 10°	01.30.39 28°	04.26.35 22°	04.44.35 20°
Mexico City, Mexico	22.05.51 41°	22.23.32 37°			

Venus transit, 2012 June 5/6

(degrees after time show height of Sun)

All times in GMT	touches edge of Sun	fully in Sun	central	exits Sun	totally out of Sun
Miami, Florida, USA	22.04.23 26°	22.22.05 22°			
Milan, Italy				04.37.37 9°	04.55.16 12°
Minsk, Belarus				04.36.59 22°	04.54.33 25°
Montreal, Quebec, Canada	22.03.34 24°	22.21.13 21°			
Moscow, Russia			01.30.30 4°	04.36.38 28°	04.54.10 31°
New York, NY, USA	22.03.39 24°	22.21.18 21°			
Osaka, Japan	22.10.47 27°	22.28.32 31°	01.29.47 68°	04.30.02 66°	04.47.36 63°
Oslo, Norway				04.36.42 15°	04.54.19 17°
Ottawa, Ontario, Canada	22.03.39 26°	22.21.17 23°			
Panama City, Panama	22.05.19 20°	22.23.06 16°			
Paris, France				04.37.23 6°	04.55.03 9°
Prague, Czech Republic				04.37.21 14°	04.54.58 16°
Quebec, Canada	22.03.29 23°	22.21.08 20°			
Reykjavik, Iceland	22.03.25 5°	22.21.07 4°		04.35.54 4°	04.53.36 6°
Rome, Italy				04.37.47 10°	04.55.26 13°
St Petersburg, Russia			01.30.06 3°	04.36.32 24°	04.54.05 27°
Salt Lake City, Utah, USA	22.05.33 52°	22.23.08 49°	01.25.35 15°		
San Francisco, California, US	22.06.21 61°	22.23.56 57°	01.25.31 22°		
San Jose, Costa Rica	22.05.31 24°	22.23.16 20°			
Seattle, Washington, USA	22.05.50 56°	22.23.24 53°	01.25.49 23°		
Shanghai, China	22.10.57 15°	22.28.46 19°	01.30.39 57°	04.30.58 7°	04.48.30 75°
Stockholm, Sweden				04.36.43 18°	04.54.19 20°
Stuttgart, Germany				04.37.26 10°	04.55.04 13°
Svalbard/Spitsbergen	22.04.32 11°	22.22.13 11°	01.28.51 14°	04.35.05 2°	04.52.41 23°
Sydney, Australia	22.15.56 13°	22.34.00 16°	01.30.15 33°	04.26.09 2°	04.44.09 20°
T'ai-peh, Taiwan	22.11.37 13°	22.29.27 17°	01.30.52 58°	04.30.39 8°	04.48.11 77°
Tehran, Iran			01.31.59 2°	04.36.49 38°	04.54.19 41°
Tel Aviv, Israel				04.37.41 24°	04.55.15 27°
Tokyo, Japan	22.10.41 1°	22.28.25 35°	01.29.31 70°	04.29.47 6°	04.47.22 59°
Toronto, Ontario, Canada	22.03.48 28°	22.21.27 25°			
Vancouver, BC, Canada	22.05.48 55°	22.23.22 53°	01.25.53 24°		
Vienna, Austria				04.37.27 14°	04.55.04 17°
Washington, DC, USA	22.03.48 26°	22.21.27 23°			
Winnipeg, Manitoba, Canada	22.04.28 40°	22.22.04 37°	01.26.01 8°		
Xi'an, China	22.10.15 6°	22.28.05 10°	01.31.06 47	04.32.14 7°	04.49.43 78°

to such an extent that it is possible to look for the small black 'Venus spot' for a few seconds without protection. The greatest care should, of course, be taken with such unprotected observation.

For qualitative planetary studies, a transit of Mercury, like that of 2003, is different in nature from a transit of Venus. Mercury shows a close alignment with the Sun in all its varied appearances so that Johannes Kepler called it the sun's Moon. Venus, in contrast, possesses a wealth of characteristics which emphasize its autonomy from the Sun. Thus its retrograde rotation indicates that the character of its motion is one that points away from the Sun. Its dense atmosphere means that it is the only planet which retains an unchanged temperature day or night — that is, independently of the Sun. Furthermore, it possesses such radiance that it can assert itself in relation to the sunlight and is already visible before sunset and after sunrise. Such an emancipation from the Sun to become something of the Sun itself is a cosmic image of what in Christian tradition is called the Luciferic principle. It also indicates, originally in the image of the 'fallen angel,' the turning away, or better cutting off, from the spiritual Sun in order to acquire inner independence and autonomy.

The transit of Mercury or Venus is qualitatively comparable to a solar eclipse because here too a body moves into the Earth–Sun axis. The fact that with a transit of Venus, only about one thousandth of the sun's surface is covered, plays less of a role here than the fact that it is not the Moon, which is turned towards the Sun in many respects, but Venus, which casts its shadow on the Sun.

2015. Venus with Saturn and Mars

As in 2010, a renewed grouping of three planets is formed five years later. Once again the assembly takes place in the constellation of Leo, but this time in the morning sky. In early October, Mars, Venus and Jupiter are uniformly distributed through Leo. This gives the gathering a magnificent emphasis. On October 9, it is enhanced further through the visit of the Moon. The image of the Moon as a vessel moving past the group of planets, drawing closer and closer together, gives an impression that the Moon is assimilating the whole of the stellar sky and offering it to the planetary community. By October 18, Mars and Jupiter are only one third of a degree apart, and one week later, Venus reaches Jupiter while at the same time Mars is positioned at the base star of Leo. The final conjunction takes place on November 3, once again a week later: now Venus has caught up with Mars so that the two neighbours of the Earth are positioned together. Now the planetary group expands again week by week. It is no longer Leo which the planets take hold of together but Virgo. In an ever growing line they enfold this large constellation in the zodiac during Advent.

In physics, the gravitational relationship of three bodies cannot be clearly grasped. Three bodies interact in such a complex way with one another that strict calculations are no longer possible. This phenomenon of physics is well known in social life. Describing the relationship of three people to one another, or, indeed, to understand it, is much more difficult, complex and intertwined than the simple relationship between two. Anyone who observes the transformation of this constellation of planets in the autumn morning sky with attentive feeling, may experience the reflection of the complexities of life when three people are together. At the turn of the year 2015–16 the constellation expands so far that Jupiter in Leo also becomes part of the series. The planets Saturn – Venus – Mars – Jupiter are therefore positioned in ascending order like a kind of celestial ladder at the start of the new year, with the Moon moving along the line of planets from January 1, and uniting them. The conjunction on

LEO

Regulus

Saturn

Mars

Venus

Mercury

Venus with Mars and Saturn in the morning sky 2015 October 9

East

January 7 of lunar sickle, Saturn and Venus next to Antares, the main star in Scorpio, forms the conclusion of this dance of the planets, which has lasted the previous six months.

SPECIAL ASTRONOMICAL EVENTS